The Trauma Controversy

SUNY series in the Philosophy of the Social Sciences

Lenore Langsdorf, editor

The Trauma Controversy

Philosophical and Interdisciplinary Dialogues

Edited by

Kristen Brown Golden

and

Bettina G. Bergo

Published by State University of New York Press, Albany

Printed in the United States of America

For information, contact State University of New York Press, Albany, NY
www.sunypress.edu

Production by Diane Ganeles
Marketing by Fran Keneston

Library of Congress Cataloging-in-Publication Data

The trauma controversy : philosophical and interdisciplinary dialogues / edited by Kristen Brown Golden and Bettina G. Bergo.
p. cm. — (SUNY series in the philosophy of the social sciences)
Includes bibliographical references and index.
ISBN 978-1-4384-2819-2 (hardcover : alk. paper) — ISBN 978-1-4384-2820-8 (pbk. : alk. paper)
1. Psychic trauma. 2. Psychoanalysis. 3. Phenomenological psychology.
I. Golden, Kristen Brown, 1964– II. Bergo, Bettina.
BF175.5.P75T727 2009
155.9'3—dc22

2008054158

10 9 8 7 6 5 4 3 2 1

Contents

Part 3
Trauma and Clinical Approaches

Part 4
Trauma and Recent Cultural History

Part 5
Afterword

Illustrations

Acknowledgments

We would like to thank, first of all, our contributors for their impressive chapters: Sara Beardsworth, Sandra Bloom, Idit Dobbs-Weinstein, Michael Galaty, Judith Herman, Gregg Horowitz, Michael Lambek, Eric Nelson, and Charles Scott. Thanks also to Sharon Stocker and Charles Watkinson. A grant from The Mississippi Humanities Council and financial support from Millsaps College played a crucial role at the volume's early stages; and Richard Smith, senior vice president and dean of Millsaps College, and the Millsaps Department of Philosophy accommodated the conference out of which this collection grew. Mary Beth Mader, John Protevi, François Raffoul, Frank Schalow and Steve Smith provided important perspectives for the volume's inception. And gratitude is owed to Eric Nelson, without whose efforts, this book would not have happened.

Michael Lambek and Charles Scott gave precious comments on versions of chapter 1, "Introduction" and insights into the production of the collection. Thanks to John Thatamanil, the State University of New York Press readers, and Jane Bunker for incisive suggestions on the manuscript, and also to members of the Millsaps Works in Progress Group and the Mississippi Philosophical Association for criticism on earlier versions of chapter 4, "Trauma and Speech as Bodily Adaptation in Merleau-Ponty." We gratefully acknowledge the following publishers and journals for allowing some of the chapters in the collection to be reprinted. Another version of Kristen Brown Golden's chapter "Trauma and Speech as Bodily Adapta-tion in Merleau-Ponty" appeared as "Nietzsche After Nietzsche: Trauma, Language, and the Writings of Merleau-Ponty," in Kristen Brown, *Nietzsche and Embodiment: Discerning Bodies and Non-dualism* (Albany: State University of New York Press, 2006), 121–49. Chapter 6, "Trauma's Presentation," by Charles Scott first appeared in Charles Scott, *Living with Indifference* (Bloomington: Indiana University Press, 2007), 125–34. Judith Herman's "Crime and Memory" (chapter 7) was first published in *Bulletin of the Academy of Psychiatry and Law*, 23, no. 1 (1995): 5–17. Chapter 9, "The Snake That Bites: The Albanian Experience of Collective

Trauma as Reflected in an Evolving Landscape," by Michael Galaty, Sharon Stocker, and Charles Watkinson is a modified version of "Beyond Bunkers: Dominance, Resistance and Change in an Albanian Regional Landscape," *Journal of Mediterranean Archaeology* 12, no. 2 (1999): 197–214, © Equinox Publishing Ltd 1999. And another version of Bettina Bergo's "Trauma and Hysteria: A Tale of Passions and Reversal" (chapter 11) appeared in Karyn Ball, ed., *Traumatizing Theory: The Cultural Politics of Affect* (New York: Other Press, 2007), 1–39. Thanks to Edith Feher Gurewich, editor in chief, Other Press, for her permission. The editors want to acknowledge Arnulf Rainer and the Arnulf Rainer Museum, Baden, for permission to reprint on the cover an image from the artist's Hiroshima series (Berlin: Neuer Berliner Kunstverein).

Kristen Golden thanks her husband Bruce Golden for his support throughout the collection's production. Bettina Bergo is grateful to the Radcliffe Institute for Advanced Study, where much of the research for her chapter was conducted. Sincere thanks also to Gabriel Malenfant (University of Iceland), Philippe Farah (Beirut), Andréane Sabourin-Laflamme (Université de Montréal), and David Bertet (Université de Montréal) for their research, criticism, and help.

1

Introduction

Kristen Brown Golden and Bettina Bergo

Recent philosophy is increasingly pluridisciplinary. Philosophers have combined existential phenomenology's approach to "life" with contemporary neuroscience, psychoanalysis, and, occasionally, aspects of social work. Interdisciplinary work has weakened parochialisms, promoting a perspectival approach to questions of trauma in their complexity. The chapters in this volume combine disciplinary influences. In one case, psychoanalysis and critical theory confront each other. In another, clinical practice is supplemented by social advocacy; in still another, ethnographic fieldwork is analyzed by cultural-formation theories. And yet, each chapter also implicitly commits itself to the assumptions and discoveries proper to its particular disciplinary expertise. Gregg Horowitz's chapter, for example, works with psychoanalytic theory; this field of inquiry, of course, draws from two principal sources: the psychoanalytic consulting room and the now vast corpus of metapsychology. While Horowitz is not a practicing analyst, he is a sensitive theoretician. His metapsychological exploration of trauma is supplemented by two practicing therapists, Judith Herman and Sandra Bloom. This is one of many textual crossings in the volume. The overlapping and dialogue between the chapters evince clarity and insight. Beyond complementarity, however, we have sought to bring differences and delineations to the multiplicity of voices in the book.

Anthologies have explored trauma from the perspectives of literary criticism and clinical psychology.[1] We find collections of anthropological perspectives on trauma in different contexts.[2] However, no trauma anthology has appeared that is both pluridisciplinary yet philosophical in its primary emphasis. *The Trauma Controversy: Philosophical and Interdisciplinary*

Dialogues fills this lack. The chapters in this volume are organized according to the following rubrics:

1. “Trauma and Theoretical Frameworks: Psychoanalysis and Phenomenology”
2. “Trauma and Bodily Memory: Poetics and Neuroscience”
3. “Trauma and Clinical Approaches”
4. “Trauma and Recent Cultural History”
5. “Afterword”

Those who have followed attentively the unfolding of trauma studies are likely aware of the enduring difficulties that have beset the field, resulting in theoretical “localities” analogous to the patchwork of a quilt. We do not presume to have gathered all the areas of this quilt; nor do we claim that the problems posed by trauma can be definitively resolved. Nevertheless, the divisions of this volume explore the crucial areas of the metaphoric quilt’s contemporary composition. Findings from divergent disciplines entail contrasting epistemologies and envision different objectives; here, many of them prove interconnected in decisive and surprising ways. Above all, it is thanks to the depth of each contribution that viewpoints arise around shared interstices. These confrontations open to novel appraisals, because the tensions between them are not reduced.

Part 1. Trauma and Theoretical Frameworks: Psychoanalysis and Phenomenology

The chapters in this first section, by Gregg Horowitz, Sara Beardsworth, and Kristen Brown Golden, evince the necessity and fecundity of pluridisciplinarity. By asking what sort of loss can be identified in the analysis of trauma, Gregg Horowitz and Sara Beardsworth offer pertinent strategies for avoiding the common conflation of trauma and loss. The fortuitous joining points that reciprocally confirm their definitions of trauma, its collapse of present experience, and its fundamental difference from sustained loss amount to an important distinction.

While loss is essential to what is called trauma, the concept has suffered undue inflation recently, with concomitant, sometimes perverse, effects: reactive channeling of grief into violence, promises of ready healing. It is therefore essential to sketch an analytic of loss itself. For the latter is tied to grief and omnipresent in childhood development. But this loss is not comparable to the devastation of trauma.

Working with Freudian theory, Gregg Horowitz examines loss as it relates to the dialectic between the dual authorities of the sufferer, whose experience only she knows, and of metapsychology, an expertise reserved for the psychoanalyst. Freud's famous patient, Dora, suffered from latent-dream thoughts, "those thoughts we spend our days and nights not believing." These are structured by a dialectic of knowing and not knowing, at whose heart, explains Horowitz, rests an internal authority. The motivating wish of the sufferer is easy to understand: "'I want my mother's love' or 'I must have my father's approval.'" Now, the problem is not that the sufferer does not know this. It is, rather, that her internalized authority refuses to believe it. Horowitz thus shows that in the psychoanalytic situation, the key to suffering lies not in knowing its source, as was long argued, but in reversing the sufferer's resistance to disclosing it to herself.

Horowitz makes a distinction between loss, which "happens to people" and is constitutive of a developmental history, and trauma, which "eviscerates the prospect of any development of a psychical structure that might measure it." By distinguishing trauma, which he calls "the persistence of the injury itself," from loss, the reemergence of a past "in and as fresh experience," Horowitz shows that with loss, one remains bound by a demand that is itself incomprehensible: not to disclose that which one understands but refuses to believe. Trauma, by contrast, is "an all-too-obvious force"—intrusive, overwhelming—that one has seen time and again. It is temporally explosive—unbinding past, present, and future and wrecking chances for experience. Loss may be incomprehensible, but unlike trauma it has available to it the temporal mediations requisite to normal development or recovery.

That we distinguish between losses that threaten the personality and losses typical to normal childhood development and the course of adult life is crucial for understanding trauma. Structural losses like that of the infant–mother bond are generally not traumatic. In most cases, they should be set apart from the persistence of injuries that disable development or everyday well-being. If the problem unleashed by mistaking trauma for a normal disruption is for Horowitz a valorizing "of violence that blunts the demand to see traumatic suffering," the problem for Sara Beardsworth is that, at the level of culture, it cloaks a forgotten loss, whose forgetting catalyzes nothing less than the onset of modernity.

Beardsworth sets the question of trauma into cultural criticism. Although Beardsworth agrees with Horowitz that the tendency to conflate trauma and loss is a mistake, she asserts that the confusion of loss and trauma is more than mere cognitive error. Her chapter carefully reveals a hitherto overlooked historical development binding loss and trauma. This, Beardsworth writes, correlates with a sense of unconscious guilt in modern subjects. It exhibits the symptoms of trauma, but is the result of a very peculiar social loss: the loss of a loss, or forgetting of a loss. Grounding her ideas in the social analyses of Freud and Kristeva, Beardsworth asks whether trauma in modern subjects is a historically conditioned structure, or an originary one innate to the human condition. In a creative rendering of Freud, she argues that the emergence of the modern subject is accompanied by the forgotten internalization of an authority whose loss is suffered unconsciously—only to be acted out.

Using Freud's metapsychology to examine the experience of the "absolute loss" (of the mythical father-prophet) that founded Jewish religion by binding the Jews to the demands of a repressed authority, Beardsworth explains that this bind was made possible by "what religion does not know, what it forgets." These developments are isomorphic with a subject whose past, present, and future are integrated, but which come undone at the threshold of modernity. A "forgetting of the loss" or loss of loss coincides with the beginning of modern subjectivity, in the aftermath of the "death of God." This loss of loss, Beardsworth argues, is a trauma whose structure is not originary, but genealogical. It is triggered by a shift, in the modern subject, away from a premodern faith in God to an Enlightenment faith in science, thus exposing a sense of guilt from a past event to which we are tied, that we have not experienced, and which will not pass.

For Beardsworth and Horowitz, effective approaches to trauma require that we distinguish suffering and depression tied to loss, from those psychic and physical disabilities that find no possibilities for discharge or sublimation. These two authors take important steps toward showing what constitutes the psychoanalytic constant called trauma.

The psychoanalysis of trauma is supplemented critically by Merleau-Ponty's approach to embodied communication. By showing that very simple organisms and simple parts of more complex bodies, such as cells, are a basis of communication, Kristen Golden reveals surprising similarities between a human body's response to physical trauma and its response to psychological trauma. Reinterpreting Merleau-Ponty's earlier ideas of human perception in light of his later ideas on communication, Golden explores the way communication is enacted by animal bodies that are structurally open to their environs and continually renegotiating and signifying their "self" and milieu. Merleau-Ponty's 1957–60 lectures show very simple animal organisms open to their surroundings. This openness, Golden emphasizes, reveals that the boundary of the "oneself" in animals is never static, but actively created in an ongoing negotiation, the very process of which makes possible a continual demarcating—a pointing to and signifying of—an inside and an outside, a "oneself" and a "beyond oneself." This corporeal negotiation of "self" and other shows the body existing as and at the root of communication. Communication happens as "interrogation (movement) and perception (response to movement)," which Golden compares with more complex communication such as human language.

By showing that communication is itself rooted in bodies that are anatomically adapting to changes in their surroundings, including drastic alterations such as wounds, Golden discloses the relevance of this model of adaptation with respect to complex communication (i.e., human language), when responding to the wound of a psychological trauma.

She draws on insights from contemporary neuroscience to show that the human body's complex systems of temporality and language, normally present in experience, actually shut down during a traumatic event. If it is worthwhile, perhaps even imperative, to create a narrative in the trauma's aftermath, then any attempt to narrate the event "truly" is not a rendering of the trauma as trauma, but a linguistic adaptation incidental to the trauma "itself." To confuse the traumatogenic narrative with the trauma itself is to misunderstand the extremity of trauma, and its contraction of anything resembling experience. Nevertheless, like a blind man who reconstructs the contours of his perceptual body to include the stick, so too the traumatized must newly create himself or herself. This often involves new practices and narratives concerning one's embodied self-regard, one's religious and ethical self, and one's self in relation to human communities. Golden's chapter roots communication in corporeity—in examples of surface wounds, neurological injuries, and traumatic violations of animals and humans. Inspired by Merleau-Ponty's investigations, she shows that psychic and physical injury share a dynamic schema: they struggle, even "want," to renegotiate the borders of "self" and milieu, both in the course of daily life and following its interruption by calamity. Although human language can be differentiated from less complex forms of animal communication, the borders between mind and body, human and nonhuman, self and surroundings are more permeable than is typically believed. On the other hand, the distance between trauma and a traumatogenic narrative is greater than generally assumed.

Part 2. Trauma and Bodily Memory: Poetics and Neuroscience

Horowitz and Golden introduced the difficult questions, Can trauma be expressed as a narrative? What sort of meaning is required in order that a narrative convey an extraordinary experience—of time (as pure intensity), other people, and circumstances? In trauma, an experience comes to pass that is so extreme that it outstrips discursive and representational resources. Nevertheless, we insist, without extensive reflection, on speaking about trauma as an experience, as though it belonged to an existential logic that could be mastered, if not by one, then by several discourses. Clearly, experience as meaningful "collapses" in cases where our ability to symbolize and represent it is severely diminished. Does that mean that, unlike loss, what we call trauma stands outside experience and cannot be justly presented in discourse, much less literature?

Many have argued that the metonymous "Auschwitz" refers to a complex, unrepresentable event: the Shoah. But "events" like Auschwitz, "Srebrenica," "Kigali"—because they surpass categories like unity, plurality, much less simple narrative time—cannot be experienced in the framework of a representational model. Further, they overflow classical phenomenology (Husserl) and its constructivist intentionality. Such events must be experienced according to a different time structure and through continuous enhancements of interpretation. But that implies that extreme trauma, in the form of historic "events," does not simply pass; it repeats and transforms itself in repeating, much the way Freud's repetition "complex" evinced a destructive plasticity.

As Idit Dobbs-Weinstein, a philosopher in our second section, has argued, "Auschwitz" is still to come; not past, not present simply, but present and to come like waves or hauntings. Sara Beardsworth showed that the twentieth-century subject lives with an obscure sense of guilt, inaugurated by an unrepresentable past. This event, which we today have yet to "experience," both passes through us and yet will not simply pass away. For Dobbs-Weinstein the repetitive and overwhelming quality of trauma provokes a host of other responses—some escapist, some sublimatory—all of which show a peculiar obsession: sustained efforts to force the unintelligible toward intelligibility. Her chapter seeks to expose—not resolve—the ambiguity of the terms "trauma" and "experience," and their relation to poetics and narrative.

Recalling her embodied childhood memory of the poem "City of Slaughter" by Hebrew national poet Hayyim Nahman Bialik (1873–1934) rather than consulting an official published version, Dobbs-Weinstein illustrates a specific or "singularly material" experience. She remembers Bialik's portrayal of a husband peeking "between the cracks to witness female relatives' sexual brutalization by the Cossacks (during the Kishiniev Pogroms of Easter 1903). Bialik's poem—or Dobbs-Weinstein's singular memory of it—describes the temptation to react to a traumatic event by retreating from its material specificity into some form of ideological or legal refuge. Dobbs-Weinstein focuses not on the brutalized women, but on the observer. Witnesses in Bialik's depictions observe a trauma, and strive to control it by escaping into religious practices or intelligible legal structures. None of these proves adequate to the event.

Through her rememoration of Bialik's poetry, Dobbs-Weinstein provides a concrete critique of those disciplines and juridico-cultural institutions that lure us to forget unintelligible experience by forcing it into everyday forms of meaning, like the language of neurosis or of simple "torts." Her work stands in opposition to over-optimistic therapeutic outlooks on trauma as a treatable "condition." It eloquently conveys the complicity and cruelty of people who "get on with the 'business of life,'" unaware of the danger and present barbarism fostered by the failure to remember Auschwitz. In so doing, they ensure that it is not past, and provide insidious nurture to the strange haunting that Auschwitz unleashed.

Dobbs-Weinstein's tone follows Adorno's, and rests on the fact that Auschwitz can and, in a sense, is happening still: people today fail to remember the singular experiences exceeding the understanding on which "all experience today depends." Above all, Dobbs-Weinstein communicates dismay at what is lost of the crime when it is translated into dominant discourses, structured by hopefulness and redemption—or brought before tribunals, whose reparative justice turns on the presumption that human experience is perspicuous. And yet, is there not intimated, in the despair expressed in poetry—or in the narrative remembrance of horror's materiality, some fragile hope, as Charles Scott seems to believe? He wonders whether Dobbs-Weinstein's bare effort to communicate does not conceal more hopefulness, at least in our material life, than, say, the promise of redemption on which the optimism of faith in a world depends.

Upon reading Charles Scott's "Trauma's Presentation," the other chapter in this section, it is difficult to imagine a position more at odds with Dobbs-Weinstein's. We have placed Dobbs-Weinstein and Scott together in this section not because of their approaches, which have little in common—Scott's argument draws on neuroscience while Dobbs-Weinstein's has recourse to "screen memories"—but because they unfold a stark polemic. Dobbs-Weinstein's reflections express steady outrage at the pervasive human tendency to deny the material specificity of horrific ordeals. Scott's chapter presents trauma as modes of indifference—to persons, and by persons to trauma. He sees in the indifference or "anesthesia" of traumatic response, a source of resilience and healthy forgetting, rather than Beardsworth's forgetting of a loss or the work of unconscious guilt evinced by Dobbs-Weinstein.

Scott and Dobbs-Weinstein agree that traumatic experience is senseless and conceptless. However, rather than criticize the modern subject's inability to attend to suffering's specificity, Scott depicts four ways trauma presents itself as a kind of indifference. One mode of trauma's indifference is an insensitivity to the injuries of others made possible by distance. "It's awful but it's not my face being shot away or anyone's in my immediate proximity." He unflinchingly describes this reaction to traumatic events when seen from afar, as in photographs and newspapers or on television.

For Scott, another kind of indifference and senselessness displayed by trauma results from physiological processes arising from trauma's disturbance to the body. When it perceives a life threat, the brain's limbic system, the seat of emotional and survival behavior, alerts neurons "to prepare for drastic action." The hypothalamus sends the autonomic nervous system into overdrive, resulting in signals conveyed "to all crucial organs, flooding the bloodstream with special chemicals and hyperactivating neurotransmitters. Respiration and heart rate increase and provide more oxygen for muscles." This survival response typically produces a discoordination between the amygdala (the seat of instinctive memory) and the hippocampal function (which converts short-term memory into long-term memory). That means that traumatic experience "knows nothing of time or place," and could not begin to do so. In light of this, trauma is neither a loss nor a symbolizable excess.

One is fortunate if the traumatic disturbance ceases with the incident. Often this is not the case. Trauma's physiology produces what Scott calls a "prereflective memory trace." One of trauma's cruelest and most threatening effects is precisely that trace, when triggered in the aftermath of the event. There it simulates traumatic experience including its own peculiar nontemporal and nonpositional orientation, reissuing the trauma as if it were present. But this is not the trauma that "was then at that place." This recapitulation joins Gregg Horowitz's psychoanalytic delineation of trauma.

In Scott's most uncanny simile, he reminds us that the somatic response to trauma in humans is like that in mice and alligators. Our limbic system's survival behavior is ancient and closer to "reptilian conditioning than human sensibility." When a traumatized limbic system dominates, whether during or after the incident, "a measure of sensation that is without [the resources of] reasonable or communal expression" fills us. This fact results in two additional kinds of indifference. The first happens in stress disorders, when intrusive memory traces appear after the event. They display the indifference to time and place discussed earlier, and which appears "in blind inappropriateness to given circumstances and in destructive noncoordination with the abilities of social consciousness." The second indifference is patent in resilience behaviors including forgetting, according to which a trauma victim cultivates an attitude of indifference to the traumatic event, and its repetition of reptilian limbic awareness. "[T]raumatic memory does not have to make a major difference in our lives," Scott argues. Despite trauma's prereflective physiology, its speechlessness, inappropriateness, Scott is not pessimistic about the ineradicability of physical memory. The physiology of trauma not only replicates the claims for the excess of trauma over discursivity, it argues for the efficacity of corporeal memory and the possibility that it can be surmounted. Emphasizing the indifference of traumatic human events to human values, Scott exhorts us not to sentimentalize or anthropomorphize trauma; as a corporeal inscription, the question of its expression and communication is not primary: there is a simplicity to trauma that is lost in its construction as an object of psychological hyperscrutiny.

Just as traumatic memory is itself "without differentiation, neutral in its disposition" and in this sense, is indifferent, "it can be forgettable and without consequence in processes of living. Its memory is sometimes expendable." Here Scott forms perhaps the most striking moment of polarity, not only with Dobbs-Weinstein, for whom forgotten trauma jeopardizes all experience, but also with Beardsworth, for whom it distinguished our entry into modernity. In concluding his arguments Scott cites Mark Twain, who "knew, I believe, as he moved on in his life, that there is a diminished future in projects that continually return to past losses."

The reptile analogy that the human limbic system offers Scott, and the material critique that Bialik's poetry affords Dobbs-Weinstein, confirm the mal-alignment implicit in the imposition of therapeutic or juridical frameworks on trauma. This attention to the ineradicable, discursive "differend" between victims' speech and legal or therapeutic discourse was not lost on Lyotard in his discussions of the burden of proof placed on victims by deniers. Nevertheless, phenomenology and psychoanalysis draw their material from everyday practice, are obliged to work with inadequate representation, and are themselves aware that trauma overflows attempts at representation. It is thus the case that both theory and practice confront starkly, if differently, the overwhelming character of trauma. In the matter of trauma "therapies," psychology and psychiatry are required to revisit certain assumptions, including that of "doing no harm," which poses a significant challenge to practitioners.

Part 3. Trauma and Clinical Approaches

Our third section presents the accounts of two pioneers in trauma therapy: Judith Herman, M.D., and Sandra Bloom, M.D. Working critically with the psychophysiological symptoms today called "PTSD" (post-traumatic stress disorder), Bloom and Herman discuss the paradox evinced in the theoretical section. For them it is of pragmatic proportions: how to work through the impasses when the will to know (that of the therapist and/or that of the victim) collides with the abyss of representation and the protean character of traumatized affect.

Herman's chapter concentrates on the problem of memory in the aftermath of a crime, describing it as a dialectic between knowing and forgetting, which leads to confusion. This confusion besets not just victims, to be sure, but bystanders as well. She compares this with Daniel Bar-On's study of children of Holocaust survivors and children of the Nazi SS. Parallels between the crime study and the memories of both groups prove striking. Notably, none of the adult children initially remembered any discussion among their families about their parent's victimization by, or participation in, mass killings. One man, whose father drove a train for the Nazis, insisted his father had never transported humans. When met with skepticism by Bar-On, his interviewer, he asked his father for more information. At their next meeting, the son reported that his father at first denied, as he always had, transporting Jews, much less having knowledge about Nazi activities. When pressed again, however, his father confessed to having been informed of these activities at the time. And then the father disclosed a tale he had never recounted before: once while on duty, he witnessed a mass shooting of war prisoners on the platform immediately before him. Bar-On interviewed the train driver's son a year later and he had no memory of his father's story of the shooting. With this example, Herman illustrates the conclusions Bar-On draws from this and many similar cases. They evince a "double-wall erected to prevent acknowledgment of the memory of a crime. The fathers did not want to tell; the children did not want to know." This disparity between knowing and wanting to know rejoins Horowitz's distinction between knowing and believing. As an illustration of the psyche's defense mechanisms, it complements Dobbs-Weinstein's censure of the "desire to subdue material experience by reason."

Herman's chapter chronicles the past one hundred years of discoveries leading to what she calls "the common denominator" of trauma: terror. In a discussion that parallels Scott's, Herman argues that trauma is the result of "intense fear, helplessness, loss of control and threat of annihilation" (citing Nancy C. Andreason), a definition also supported by studies in the *Diagnostic and Statistical Manual of Mental Disorders: DSM IV*. Terror, she explains, is an altered state in which the conditions for perceiving, paying attention, and arousal shift dramatically. People in a state of terror lose awareness of time and "peripheral detail," but become fastened on "central detail"; with the narrowing of attention comes dissociation and "profound perceptual distortions including insensitivity to pain, depersonalization, derealization, time slowing, and amnesia."

Scientific inquiry into, and public openness to, ideas about trauma have flowed unevenly, Herman reminds us, like sets of waves arriving with each major war, forcing the issue; only to recede again, the backwash giving way to scientific or grassroots backlash. With each generation of war veterans, she writes, psychiatrists reencounter the same lesson: when the survivor sets the indelible trauma images and sensations into narrative, their perniciousness dissolves. She credits the women's movement of the 1970s and 1980s for bringing to public awareness the breadth of domestic violence, incest, and rape and for framing these as human rights violations with direct parallels to political crimes that "perpetuate an unjust social order through terror."

A longtime advocate for victims of violence, Herman describes the mid-1990s as the height of controversy over the accountability of perpetrators. Research about adult recollection of childhood memories and a spate of legal cases about incest and rape, based on the integrity of recovered memories, emerged as a heated debate for American citizens, therapists, and scientists alike. Since then, more has been learned about standards for verifying adult recall of childhood memories in the therapeutic consulting room. Nevertheless, divergent discourses must be acknowledged: standards of evidence for clinical encounters, scientific research, and juridical courtrooms need to remain distinct. This does not make it easy for the therapist or bystander when called on to stand as witness in the courts. Like most persons, writes Herman, "we are not very brave" and would "rather live in peace." And "like the son of the man who drove the trains in wartime, we are reluctant to know about the crimes we live with every day."

The writing of this introduction coincides with the second anniversary of Hurricane Katrina. For every successful event that takes place in New Orleans, like the 2007 New Orleans Jazz and Heritage Festival, there are countless other stories of outrage and neglect. When Kristen Golden visited the city recently, she saw a man wearing a T-shirt that said "Screw Iraq. Save New Orleans." Expressing a local sentiment of desperation and anger, the slogan indicated a perception that U.S. leaders and the country's general population have largely ignored the city's ongoing plight, all the while keeping their attention on Iraq and funneling untold dollars toward the war effort. If New Orleans schools are any indicator, it is apparent how far the area needs to come in restoring its health. Two years after the horrific flooding, "less than half of the original public and private schools have re-opened."[3] In the most impoverished areas blasted by Katrina, the percentages are much lower. However, despite the magnitude of the Superdome and Convention Center crisis, now years ago—and the daily ordeals of those trying to live in New Orleans in Katrina's aftermath—many across the nation have effectively "gotten on with their business." If, early on, many were caught up in the dialectic of "wanting to know while unable to believe," as with so many social disasters, the pendulum of public sentiment has swung decisively toward Horowitz's "disbelief" and Herman's "wanting to forget"—even as the event persists, like an imperfect tense, like a future. As philosophers Jean-François Lyotard and Jean-Luc Nancy have done, Herman urges therapists and bystanders to envision the discursive enactment of trauma as a moral and aesthetic call, however difficult, and to "accept the honor of bearing witness and [to] stand with [patients]"—despite personal and institutional frustrations.

Sandra Bloom is also a psychiatrist and a longtime advocate for victims of violence. Her chapter begins with a compendium of statistics such as: one in two Americans will undergo an event widely experienced as "traumatizing"; 25 percent of those traumatized will develop PTSD; 50 to 70 percent of persons who are raped (and one among every eight American women are) or physically assaulted (one among every two women are) will sustain PTSD. Given the pandemic exposure of Americans to traumatizing experiences, Bloom asks why more attention is not paid to trauma, which opens a parallel between individual and social psychology. She hypothesizes the likely interconnection between individuals organized by their traumatic experience and larger social contexts (groups, political events, institutions of employment); that is to say, the probability that these become replicated on a broader scale. These "trauma-organized systems," she indicates, reenact "for individuals the very experiences that have proven so toxic for them." Is there too much speculation involved in mapping how trauma affects individuals onto broader social contexts? The question was put to Freud, as Beardsworth is aware. Nevertheless, given the pervasiveness of trauma in our culture, it is clear that insights emerge when we view social groups through Bloom's lens of "trauma systems." She argues that not to pursue the parallel is like ignoring "an elephant in the middle of a small room."

In a fascinating account of the evolutionary biology of the human response to stress, Bloom shows that the benefits it provided humans over most of our evolutionary existence now appear as a source of great harm to us in our recent species history: an industrialized global society. Using paradigms of neuroscience and human stress studies similar to those provided by Kristen Golden, Charles Scott, and Judith Herman, Bloom discloses further aspects of a nonvolatilizable core to trauma; notably, the action of the limbic system, responses from the autonomic nervous system, and the cancellation at a physiological level of time-consuming activities such as speech and reasoning, even of the temporality of the body itself. Hyperarousal is an evolutionary adaptation that protects even as it isolates persons, and perhaps groups, from therapeutic communicative options.

Bloom assumes that all human systems, whether of individuals, small groups, or nations, are structured by parallel trauma processes. In times of cultural stress, decision making and political leadership adapt their forms according to what resembles the physiological fight-and-flight response. Rallying around a figure who appears to be decisive or strong, the group allows decisions to be made more unilaterally, experiencing "pressure . . . to conform to standards of cohesion." Bloom then argues how, in stressful situations, the strategies devised to benefit a group become systems that simulate the rapid, prereflective response behaviors seen in hyperaroused individuals. To illustrate this, Bloom examines the social and political trauma around the World Trade Center's destruction and its aftermath. Defining the post–9/11 context of the American population besieged by political rhetoric as a "trauma-organized system," Bloom shows how the proliferation of media images and hyperbole not only fostered mass fear, but aborted the difficult process of grieving, and thrust it into a two-stage war that served the economic interests of a minority.

Bloom's examination of the manipulation of discourse and images, which foreshortens the work of grief, returns us to the question of the overdetermined nature of trauma and the search for intelligibility raised by Horowitz, Beardsworth, and Dobbs-Weinstein. While Bloom's observations preceded debates that have since become more widespread, her parallel examination of personal and public trauma, and the relationship between mourning and the manipulation of our urge to escape its suffering, are pertinent.

The question of trauma's social ramifications and its use by political and economic institutions also poses the question of the determinations of trauma itself as an "epistemology." There is little doubt that trauma has multiple cultural determinations. But are we entitled to argue that the very concept of trauma is itself determined by contemporary Western culture? If we take a cultural-anthropological tack (itself a creation of the nineteenth-century West), we must acknowledge the contemporary growth of a symbolic-cultural industry of trauma. Does that justify arguments that something like a trauma-constant evaporates when we recognize the relativity of cultural symbols and practices? Is there no core to trauma, then? Horowitz already cautioned us against inflating trauma, showing how we can avoid mistaking developmental loss for traumatic loss. Scott warned against a drive to symbolize that overlooked the adaptability of the body. What insights can we gain from cultural anthropology? In the first place, it is attentive to the social varieties of response and sublimation, not to mention their limits and specificity. But cultural anthropology, beyond its attentiveness to the specific societal conditions that mediate traumatic experience, confronts the question of how to think that to which diverse societies are responding. If there is no eluding the paradox of discursive reenactment of traumatized affect, then there is no escaping the necessity of symbolization and restructuring of groups' traumatic experience.

Part 4. Trauma and Recent Cultural History

Some anthropologists make the opposite claim. A conceptual and disciplinary monolith, trauma must be dismantled. Yet contemporary discourse about physiological "memory" and arguments for extradiscursive, yet psychic, memories are the products of late capitalist cultural constraints. Their source calls for critiques of the kind Foucault addressed to psychiatric practices and institutions. Anthropologist Michael Lambek argues that the "experience" of trauma makes little sense outside this particular cultural configuration. If North America stands as the apogee of medical and objectivist constructions of traumatism, the deployment of North American discourses creates insuperable tensions when imposed on cultures whose traditional practices are delegitimized in the process. The violent histories that form the context of many anthropological endeavors oblige anthropologists today to approach "trauma" from a critical and nondeterministic perspective. What would a logic, broader than that afforded by trauma as a medical or therapeutic object, require to study the lived experiences, say, of Cambodians who survived the Khmer massacres?

Archaeologists Michael Galaty, Sharon Stocker, and Charles Watkinson show that the repeated invasions, politico-cultural hegemony, and collective suffering that characterize recent Albanian history offer us an example of resistance and partial healing. While the Communist Albanian Party of Labor under Premier Hoxha essentially refashioned Albania's landscape architecture, with an uncanny proliferation of more than seven hundred thousand concrete bunkers reinforced by layers of steel, they did not foresee that these bunkers would be reappropriated to form the most apparent symbol of an Albanian recovery of the past. The spontaneous popular transformation of the mushroom landscape includes the preservation of memory and a surprising reflex toward humor. Bunkers have been turned into cafés, others demolished, while still others are preserved as miniature museums.

With the bunkers, the authors also analyze a large, eighteen-centuries-old Roman-period block. They trace the block's architectural and symbolic history from its origin as part of a Roman-period *bouleuterion* in Apollonia, a Greek colony in ancient Illyria (modern-day Albania), to a position of prominence in the nineteenth-century church of Shëndelli in the Shtyllas valley. It was rediscovered, fortuitously intact amid the rubble, when the church went the way of most churches and mosques in Albania in the 1960s and 1970s: razed by local communist parties. By the time Galaty, Stocker, and Watkinson visited the site of Shëndelli in 1998, the block had become the centerpiece in an outdoor sanctuary for worship and religious practice by Christian and Muslim villagers living in Shtyllas, a mile away. Many of the villagers were related to or themselves political dissidents forced by Hoxha to resettle in this rural area. The block's heritage, connecting its Roman and then Christian ancestry with religious practices outlawed by Hoxha, combined to make it a symbol of popular survival. Galaty and his coauthors suggest that the block has "become a key element in an invented local tradition, one that joins the past to the present and, in the context of resistance, refers back to local, rather than central, systems of government." Like the bunkers today, the block, surviving significant mutations in tradition, has become a material symbol of a societal history of trauma and resistance.

On the basis of fieldwork, Galaty, Stocker, and Watkinson have tried to ascertain how much these symbols preserve cultural memory. They show the efficacity of adaptable architectures to combine resistance, memorialization, and irony. Is this not a unique cultural therapy directed toward the undeclared condition we deem "trauma"? The multilayered symbolic status of the bunkers and the block—two examples of mnemonic "emblems" scattered across cultural geographies—is in transition. Everything suggests that the transition is not gratuitous. But to what extent can they be said to be monuments to a collective traumatic experience? How can their evolving status be characterized justly if anthropologists, theorists of diverse stripes, and therapists insist on a North American model of trauma and determine the appropriate measures for its "redress"?

Indeed, many remain unaware that a philosophy contemporaneous with the greatest ethno-social cruelty in the twentieth century contains elements influenced by early trauma theorists. Reviewing the context of Heidegger's complicity with National Socialism, Eric Nelson shows that Heidegger's 1935 *Introduction to Metaphysics*—the standard readings of which agree it endorses, or at least provides the philosophical resources for promoting, a policy of social violence—requires a more complicated reading. Nelson first outlines Heidegger's unpardonable alliance with National Socialism at its apex between 1933 and 1935. But he argues that the lectures of *Introduction to Metaphysics*, which were given after Heidegger resigned from his Nazi appointment as rector, mark a hitherto overlooked turning point in his thinking about violence. In writings such as *Beiträge* (*Contributions to Philosophy*, 1936) and others from the later 1930s, Heidegger's philosophy attempts to take distance from the values of National Socialism. Recoiling before the German drive to war, he decries the "biologism, giganticism, racism, worship of power, [and] frenzied commitment to the total mobilization of society."

Nelson reminds us that the theoretical source of National Socialism's momentum were the ideas of Ernst Jünger and Carl Schmitt, the basis for which was a romanticizing of the "traumatized life." Jünger valorized soldiers who had endured traumatizing experiences under fire, during World War I, and envisioned mobilizing the masses of an entire nation based on the soldiers' example, as having existed in a state of constant mobility, homelessness, and threat to their lives. Jünger's vision parallels Bloom's social psychology discussion of traumatic systems: their unbridled reenactment of physiological mechanisms of shut down and response, and unwitting simulation in larger organizations. With Jünger, the replication is romantic, and for us, unacceptable; but it points to the impact on philosophy and literature of large scale traumatization.

Schmitt, who became the Third Reich's leading jurist, theorized creating a state of emergency on a national scale by hyperbolizing the external and internal threat of enemies. This macrocosmic parallel to the terrorized individual facilitates the rise and acceptance of an overbearing leader. Like that which Bloom mentions, the emerging leader, preeminent symbol of protection, is permitted to make rapid decisions for the populace, conveniently circumventing democratic deliberative processes.

If Bloom has grave misgivings about the adequacy of our evolutionary stress responses—adaptive mechanisms with an astounding success story for tens of millennia—to do much else, in an industrialized global context, than harm us when surreptitiously stimulated in large groups, Heidegger, Nelson suggests, had increasing doubts, through the 1930s, about the compatibility of his reflection on violence and that of the National Socialists. By divulging a generally overlooked aspect of Heidegger's view of *polemos* in Heraclitus and Nietzsche, Nelson shows Heidegger conceiving strife as an originary mode of human being. If the source of the sense of alienation and overwhelming rests in the experience of the human individual, then Heidegger provides the "basis for a critique of the self-assertion of egos and races." By indicating that the source of our unease is constitutive of human existence per se, Heidegger "throw[s] into question" the motives for intrahuman conflict and positions his reflections on violence in opposition to Schmitt's, for whom conflict between human groups "is always justified as the essence of the political."

Nelson's striking reading of Heidegger's view of Nietzsche's *On the Genealogy of Morality* as a genealogy of trauma highlights Heidegger's growing rift with the National Socialist agenda. National Socialism's response to violence is like Nietzsche's life-denying priest, who fails to respond to the wounds of violence or trauma, except to reinscribe and deepen them by leaving them "unencountered and unquestioned." For Nelson, Nietzsche's genealogy traces the transformations of trauma, or one might say a series of traumatic origins. One such traumatic origin is human beings coming to understand themselves as things. In this sense of self-reification they see themselves constituted by a certain unchanging rational nature. After citing a history of torture and cruelty that makes humans into calculating beings, and the legacy of revenge and resentment convoluting trauma, violence, and love, Nelson cites Nietzsche's "gruesome paradox of a 'god on the cross', that mystery of the inconceivable, final, extreme cruelty and self-crucifixion."[4] For Nietzsche, the permutations of suffering and trauma, Nelson argues, are rooted in human "practices and institutions." Thus, Nelson joins Beardsworth's social genealogy.

The celebration of the human "traumatized life" by Ernst Jünger and Carl Schmitt presupposes self-reification as a cultural practice, something that Heidegger's philosophy of being—typified by uncanniness, strife, and dispersion—ultimately put into question. Nelson argues that, in the 1930s, Heidegger was searching for ways to "let the wound appear as a wound." He was asking philosophically, "Could being wounded call forth a response that recognizes its wounded character?" The blind reactions, specific to many violent behaviors, appear perpetrated against a falsely constructed enemy by a falsely reified self. According to Nelson, if Heidegger's conception of existence is characterized by anguish and conflict, overwhelming a coherent human subject, then Nazi war and genocide, made possible by imagined selves and enemies, exposed Heidegger and others' failure to respond appropriately to the strife following World War I and the demise of Weimar. Nelson's study brings to light the reactivity that Heidegger's philosophy embodies, and then questions. Now, while Nelson and Beardsworth link trauma to social and institutional developments dating back centuries, Bettina Bergo focuses on its emergence in the recent history of psychoanalysis.

Some ethnographic epistemology relativizes psychiatric and psychological aspects of trauma. Part of the self-critiques of Western theories of experience and knowledge, it negotiates a path between a suspicion about trauma as experience *eo ipso* and a critical nuancing of trauma as an epistemic object. Above all, it is not clear that the ethnographic critique dissolves the possibility of trauma having a dual, psychic and physiological, core. Because cultural practices exert a determinant impact on the interpretation of experience and its transmission, understanding this impact calls for a critical history of psychoanalysis and psychiatry. Bettina Bergo's study of Freud's evolving approach to trauma as hysteria complements the ethnographic chapters by adumbrating Freud's own struggle—first to dissociate hysteria from gender and anatomy and place it within a traumatogenic context; thereafter, to distinguish hysteria from war trauma. Freud's deliberations were based on his work with Charcot at the latter's veritable trauma factory, the Hôpital de la Salpêtrière in Paris. It was Charcot, then Freud, who first revealed the cultural conditions sufficient for trauma to appear in women and in men as hysterical paralyses, aphasia, and lesionless "epileptic" attacks. Bergo argues that the contemporary diagnosis of PTSD resulted from historical transformations in the medical conception of male and female physiologies. Paradoxically, Freud gradually left hysteria behind him as a disorder tied to traumatism to return ultimately to more traditional gender distinctions following World War I. The strange career of the nineteenth-century condition called "hysteria," the displacement of traumatism (in women) from wombs to memories (in men and women) fostered a thirty-year debate around the differences between neurosis, "shell shock," trauma, and genders. In so doing, it had transformative repercussions first on the Central European cultural imaginary, and thereafter on the discourse of contemporary psychiatry.

Bergo's chapter shows trauma requiring that certain cultural ideas about human psychology and biology be in place in order for it to appear conceptually and experientially. If this is true, then what can we learn about trauma from the mutations in its interpretations?

Part 5. Afterword

The collection has brought to light three major themes, which it approaches pluridisciplinarily. First, it presents and evaluates arguments for the persistence of a dual core in trauma; that is, a physiological and psychological core. Criticism from ethnographers and psychiatrists working with trauma victims raise questions concerning the social variations in and translations of this traumatic "core." Second, despite difficulties in the construction of trauma as an object of knowledge and clinical practice, the chapters show that there is a dialectic between individual and social trauma. This dialectic cannot be reduced to developmental or cultural experiences of loss. Third, the volume presents arguments against the appropriation of trauma by discourses of redemption or reductive impositions of "meaning"; some chapters show how these blunt the extremity of trauma. Nevertheless, without psychologizing literary expressions of or philosophical approaches to it, we can study symbolic responses and compare cultural sublimations of trauma.

The afterword by Michael Lambek critically reviews each of the chapters in the volume from a cultural-anthropological analytic of discourse. Taking a broader stance than either Bloom or Galaty and his coauthors, Lambek investigates the conditions under which scientific and literary ideas unfold and spread. In his view, a concept and experience of trauma can hardly make sense independent of a particular cultural configuration, North America's, for instance, with its scientific, medicalized, and objectifying approach to suffering. "Objectifications, like Frankenstein's monsters, sometimes take on a life of their own and may even contribute to the very effects they were designed to suppress," he writes. It was in order to check such a "Frankenstein effect," that we invited Lambek to write the volume's postscript. Lambek's fieldwork in Madagascar has shown how memory, often traumatic, is a cultural and moral practice. Its great interest lies in the flexibility with which Lambek moves between ethnographic specialization and epistemologies of practices in many cultures.

Perhaps the most striking of his commentaries comes in a story of Canadian aid workers in northern Uganda. Lambek describes workers "busily diagnosing and treating PTSD among children" who had been torn from their families and forced to participate in brutal massacres. The irony is that PTSD presupposes a return to social conditions whose "normal" state is stable and secure. While stability is culturally important, such a therapy is irrelevant to cultures caught in cycles of civil war and economic disintegration. By concealing the unrest, do aid workers unwittingly "collude in its effects and possibly even with the structural forces that lie at its origins?" Lambek underscores the looping effect that occurs between a subject being "kinded" (i.e., categorized as "traumatized") and what a subject believes about how he or she is treated due to being so kinded. Wary of literal "trauma," Lambek urges theorists and practitioners of trauma to ask how a discourse of "trauma" spreads. Combined with Bergo's historical approach, Lambek's argument shows that, without a secular, medicalized conception of suffering in therapy—and a cultural imagination capable of receiving it—PTSD could not congeal into a class of disorders set apart from its symbolic context.

The volume thus begins with Gregg Horowitz specifying trauma, not to reify "it" but to explain its distance from developmental losses. It closes with

Lambek, skeptical about unilateral historic and epistemic categorization. Our purpose throughout is to open unanticipated dialogues between disciplines and encourage stances that encompass conclusions drawn from psychiatry, philosophy, ethnography, and therapy.

Notes

1. See Linda Belau and Peter Ramadanovic, eds., *Topologies of Trauma: Essays on the Limit of Knowledge and Memory* (New York: Other Press, 2002); and Cathy Caruth, *Trauma: Explorations in Memory* (Baltimore, MD: Johns Hopkins University Press, 1995).

2. See Ana Douglass and Thomas A. Vogler, eds., *Witness and Memory: The Discourse of Trauma* (New York: Routledge, 2003); and Paul Antze and Michael Lambek, eds., *Tense Past: Cultural Essays in Trauma and Memory* (New York: Routledge, 1996).

3. Amy Liu and Allison Plyer, "A Review of Key Indicators of Recovery Two Years after Katrina," in *The New Orleans Index* (New Orleans, LA: The Brookings Institution Metropolitan Policy Program and Greater New Orleans Community Data Center, 2007), 13.

4. Friedrich Nietzsche, *On the Genealogy of Morality*, trans. Maudemarie Clarke and Alan J. Swensen (Indianapolis, IN: Hackett, 1998), 1:8. The first number refers to the treatise, the second to the section.

Part 1

Trauma and Theoretical Frameworks: Psychoanalysis and Phenomenology

2

A Late Adventure of the Feelings

Loss, Trauma, and the Limits of Psychoanalysis

Gregg M. Horowitz

Metapsychology and Psychotherapy

To the unending chagrin of its practitioners and proponents alike, psychoanalysis remains an incompletely unified discipline.[1] On the one hand, it offers a metapsychology, an integrative theory of the mind that is related to psychology, the science of the mind, much as metaphysics is related to physics. Psychoanalytic metapsychology deals with problems that are generated by psychological inquiry but that demand sustained theoretical and speculative reflection even after the scientific fruit of psychological inquiry has been harvested. Psychoanalytic metapsychology, in other words, is not directly productive of scientific knowledge, but rather serves to bring to reflective attention the cognitive work that remains even after the scientific work of understanding the mechanisms of mind is finished. This is why metapsychological texts so often seem uncanny doppelgängers of philosophical texts; like philosophy, psychoanalytic metapsychology is a field of inquiry that takes form because empirical knowledge of the mind is not complete and self-sustaining. As such, the capacity to produce metapsychological concepts for knowing the mind is constrained not merely by exigencies of psychoanalytic experience and, more generally, the empirical facts of psychology, but also by the norms of metalevel reflection on the aporetic nature of mindedness as such.[2]

The demands of metascientific reflection are strenuous enough that on their own they problematize the prospects of a final, consistent, and unique metapsychology. When I referred to the incomplete unification of psychoanalysis, however, I had another source of disciplinary dissonance in mind, for while psychoanalysis is a metapsychology, it also remains a therapeutic practice. As such, its aim is to relieve disabling emotional and psychical distress, to return to diseased minds the capacity for healthy engagement with the difficulties of life. It is certainly true that the therapeutic techniques for relieving pain and suffering characteristic of psychoanalysis are so highly particular that they sometimes strike us more as discourse than as doctoring, but that is only because the talking cure tailored to pathologies of the mind, shot through as it is with all the vicissitudes of speech and audition, of address and misrecognition, is a close kin of nontherapeutic practices like confession, philosophy, and hypnosis. But as a therapeutic practice, psychoanalysis, its filiation from nontherapeutic practices notwithstanding, is a medical discipline, a phronetic discipline, driven by the demands made on it by the suffering of patients. Indeed, one might reasonably assay, as did Stephen Mitchell and Margaret Black, to understand the historical development of psychoanalysis by examining the history of the presentations of pain that have demanded therapeutic response from it.[3] This current chapter is in the spirit of Mitchell and Black's exercise, albeit with a twist, for it is my view that the specificity of traumatic suffering places not just formative demands but also deformative demands on psychoanalytic theory and practice.

Regarded as a therapeutic practice, the purpose of psychoanalysis is its capacity to relieve pain and suffering. Therefore, its success is measured not merely by the facts of empirical psychology but also by the norms that govern healing. Psychoanalysis is, therefore, a discipline that answers to two masters, one metapsychological and the other medical and practical, one conceptual and the other experiential and intersubjective, one reflective and the other highly specific and concrete. The relation between these two different aspects of the discipline is always a tense one, insofar as the two aspects embody different cognitive and practical aims and aspirations. Sometimes, however, psychoanalysis' two masters issue demands that are not merely refractory but intractably incompatible. For example, while theoretical and conceptual consistency at the metapsychological level is a measure of the correctness of the reflective ordering of clinical experience, and so of the communicability of that experience among analysts, it is not on its own a criterion of clinical and practical effectiveness. Indeed, excessive insistence on metapsychological completeness and consistency in the face of anomalous clinical presentation is the sign of a doctrinal rigidity from which inevitably follows poor hearing, or even nonhearing, of the clinical presentation. In this light, it is significant that the first case published independently by Sigmund Freud, "Fragment of an Analysis of a Case of Hysteria," the famous primal case of Dora, was the story of a failed psychotherapy. Freud's treatment of Dora was marred by his insistence on an etiology of Dora's suffering as deriving from her inability meaningfully to say "no" to the predations of her putative guardians. The problem in the treatment of Dora was not that Freud wrongly diagnosed the etiology of her illness, but rather that, in authoritatively insisting in word and deed on his rightness, he ended up preventing Dora from saying no to him and thereby hounded her out of his consulting room. The doctor's obedience to the demand for metapsychological consistency and correctness in the face of the presentation of the patient's suffering blocked all prospects of her working it through. It was a practical disaster for the patient.

Freud published his history of the Dora case not merely despite the failure of the treatment, but in large measure because of that failure. Dora made demands on her doctor that the doctor did not hear; "I do not know what kind of help she wanted from me," Freud concludes in the penultimate paragraph of the case history.[4] In presenting Dora's case as a lesson in the consequences of not mastering the transference—I do not know what kind of help she wanted *from me*, Freud says, thus disclosing that he did not know who he was for Dora—Freud explains that he could not hear his patient's demands because he assumed they were demands on the person he took himself already to be—the doctor, the medical authority, the stranger to the drama of Dora's family life—rather than demands on figures operating darkly inside of Dora's phantasies whom she was addressing, in their physical absence, there in the consulting room. In taking Dora's demands at face value, Freud thereby let his metapsychological understanding govern his audition of Dora's presentation and so allowed himself to merge with the paternal authority against whom Dora's protests had long been in vain. He ended up, therefore, not knowing, but the not-knowing in question was generated out of too much knowing, or at least too much knowingness, about what the objects of Dora's desires were—about what, as he puts it, she really wanted. What followed from Freud's overemphatic commitment to theoretical consistency was an appropriation of narrative mastery over the patient's desire that inhibited the analytic articulation of that desire, and what followed from that was a chain of gross clinical errors. The lesson learned from the case history of Dora can, then, be put this way: excessive commitment to the metapsychological authority of psychoanalytic concepts can serve to protect the analyst from hearing the illicit demands of the patient and, thus, to keep those demands from the therapeutically necessary confrontation with the authoritative narrative forms that shape and distort their expression.

This insight into the inhibitory function of metapsychology generates the need for a critique of psychoanalytic knowingness. As Jonathan Lear has pungently observed, psychoanalysis nowhere betrays its promise more utterly than when it presents itself paternalistically as doctrine and method that can authoritatively bring to conscious awareness all the determining forces of character.[5] Psychoanalysis' primal scene, Lear reminds us, is the confrontation between insight and paternal authority in which paternal authority displays its uncanny, automatic ability to trump all merely intellectual rebuttal of its insistent, mechanical demand that one sacrifice one's desire to it. When Freud discovered that convincing a patient of the truth of an interpretation can just as easily further sicken as help the patient, he realized that the aim of psychoanalysis had to be not merely to achieve Socratic self-knowledge but rather to strengthen the psyche in its combat with an authority that can sometimes appear precisely in the form of the demand to know yourself. The reflective understanding of psychical conflicts that constitutes the psychoanalyst's theoretical expertise is thus no substitute for the patient's self-appropriation of the persistence of those conflicts, and the conviction that it can be a substitute must be regarded as a disciplinary deformation. From this point of view, the encounter with Dora is the scene of the violence that the psychoanalyst does to a patient when he refuses to allow his metapsychological concepts to be disturbed by her clinical presentation. To call into question the metapsychological handling of the problem of analytic authority does not promise a more solid ground for that authority. Rather, it opens psychoanalysis up to radical uncertainty about the function and value of authority. Psychoanalytic self-criticism begins, then, with an "insight" into both the limits of the force of metapsychological cognition—knowledge will not, after all, make one free—and the danger of forgetting those limits.

Unfortunately, however, this critical insight into the dangers of authoritative knowingness, its crucial importance notwithstanding, can also lead to dangerous distortions in its own right. (No insight, it seems, arises unaccompanied by its own special power to damage.) The analyst Roy Schafer, who did so much to bring the results of critical narratology to bear on softening the scientistic bent of postwar American psychoanalysis, recently delivered a lecture in which he found himself entertaining the thought that in the encounter between patient and therapist the therapist brings to the treatment no authority whatsoever.[6] As Adam Phillips has put a similar view, in regard to living a life, no one is an expert.[7] Now, while we can appreciate Phillips's humane aphorism, the practical conclusion it points to is incoherent in the analytic context, since without authority being brought into the consulting room, there is no beginning to psychoanalytic practice. Indeed, because the patient brings with her into the consulting room the expectation that the analyst has an expertise that promises the possibility of therapeutic results—the expectation that the analyst is, in

Lacan's phrase, the subject-supposed-to-know[8]—we must suppose that the premature shedding of authoritative standing by the psychoanalyst may be just as damaging as the defensive claiming of it. Notwithstanding that it must be seen as a self-critical overreaction, the derogation of metapsychological authority is nonetheless (as my citations have shown) a pattern of thought to which a significant body of contemporary analysts are drawn in the face of their experience of the historical misalliance within psychoanalysis between knowing too much and hearing too little or, putting the point in a theoretical register, in the face of the breakdown between metapsychological knowing and the healing of pain. In this self-defeating dismissal of analytic authority—on presumably authoritative grounds—the war between the metapsychological and therapeutic elements of psychoanalysis comes out into the open. And because the conflict breaks out at the site of the secret pact between authoritative knowledge and authoritative suffering, the direction of a truce is hard to detect. Without going too far into the state of recent debate about the authority of psychoanalytic knowledge, I think it is licit to say at least this much: the disunity at the heart of the discipline of psychoanalysis comes from a persistent uncertainty in its conceptual relation to its practical work and its practical relation to its conceptual work. Some profoundly broken relation between cognition and suffering is at work at the center of psychoanalysis.

There is, to be sure, a way for psychoanalysis to slice off one horn of its dilemma. Under various institutional pressures, the therapeutic practice of psychoanalysis can be farmed out to pragmatically or pharmaceutically inclined practitioners of shorter-term therapies who are equipped with steady and practicable criteria for cure; at the same time, the metapsychological element can become the property of academic critical theorists. Arguably, this is, in fact, the direction in which the field as a whole is currently shifting; in consequence, we find ourselves in the kingdom of universal eclecticism where sketchy clinical results give rise only to alternative therapies, and critical theory liberated from clinical practice yields nothing but theoretical speculation and endless conceptual refinement. This path is, I believe, mistaken for both the therapeutic and metapsychological aspects of psychoanalysis. The belief that the tension between the demands of suffering and the metapsychological inquiry into the cognitive adequacy of our concepts of mind can simply be dissolved rests on the assumption that there is no intrinsic connection between the suffering that psychoanalysis attends to and its forms of knowing. It assumes, in other words, both (1) that the demands that give suffering its human texture can be addressed otherwise than through the mediation of knowing and (2) that cognition can find orientation otherwise than through its relation to suffering. These assumptions represent, however, just the unbinding of the secret pact between knowledge and suffering, which it is the special responsibility of psychoanalysis to preserve. Understood as the unstable conjunction of metapsychology and clinical therapeutics, psychoanalysis is based on—more strongly, it is generated and bound by—the aporetic insight that while knowledge on its own will not emancipate us from the suffering of our proper histories, nothing else will either. This, I believe, is the ethico-political thrust of psychoanalysis understood as a critical metapsychology with an emancipatory intent. To evade the psychoanalytic thought that the persistence of need in cognition is both a limit of and a condition for cognition's curative force is to fail to heed the critical idea that need persists within—which is to say, is suffered by—conceptual work in the form of incomprehensible demand. The fate of incomprehensible demand, of the corporeal, insistent, material residue of personal and public histories within current psychical life, is the stake to which psychoanalysis ties emancipation from suffering. And with this claim about the binding significance within cognition of what remains uncognized, I finally approach my topic for this chapter, for it is in the confrontation with incomprehensible loss and traumatic suffering that psychoanalysis generates the specific ethico-critical force that troubles metapsychology.

Patent Loss

What does it mean to call loss, or anything else for that matter, incomprehensible? One common understanding of incomprehensibility regards it as a species of the mysterious, as what is hidden or withdrawn from view. Incomprehensibility would be, in that light, an attribute of an unknown bit of reality that beckons to cognition to conceptualize it otherwise. My worry about yielding to this common understanding is that it tempts us, in the psychoanalytic context, to conceive of suffering as such as incomprehensible. Suffering so conceived would be suffering already conceived, and suffering already conceived is, as I suggested in my sketch of the confrontation between psychoanalytic knowing and suffering, suffering evaded. If we are to grasp the specificity of incomprehensible suffering without reifying it, we need to resist unifying all suffering under a single concept of incomprehensibility. We need, in other words, to make a distinction, however strange this sounds, between variants of incomprehensibility. That is my aim in the following sweep of argument.

Let us start by noticing that there is, encoded even in common speech, an alternative to the positive characterization of incomprehensibility. We often respond to a narrative of suffering by saying that what the narrator has gone through is unbelievable, incredible, or unimaginable to us. In saying so, we acknowledge the significance of the narrated experience by avowing the limits of our capacity, not so much to conceive the suffering but, rather, to credit it. In such cases, we suffer no deficit of understanding, but we nonetheless experience a persistent demand on us even after we are done understanding the story. Whatever in the narrated experience remains incredible—whatever cannot be credited to the auditor's experience—has nonetheless been heard and, if the story is well-articulated, understood, but the meaning of undergoing that experience from the first-person point of view remains ungrasped. The tale may be clear, but the telling remains murky. We might even say that the inability on the part of the narrator to communicate the experience even within a perfectly well-formed story is the compelling mark of its being a story of suffering. There is no mystery in such cases about some specifically incomprehensible moment of suffering, but rather a lack of experience with which to measure the suffering.

Another way to put this point would be: there need be nothing deep about the incomprehensible. While in the moment of telling and listening we often treat what eludes our belief as if it possessed a uniquely powerful positive charge of incredibility, it often turns out that what we cannot credit at a certain moment reveals itself to be simple and easy to believe when we are not defended against it. In such cases, "incomprehensibility" is a property of our relation to something that, despite being easy enough to grasp in fact, we nonetheless cannot bring ourselves to believe. The deficit of the "in-," is not, then, an epistemic one. We know what the source of the suffering is, but what we do not get yet (and what we point to with the concept of incomprehensibility) is how it is a source of, specifically, suffering beyond our knowing—of a suffering we cannot overcome merely by further cognition. We know the source of the suffering, but whatever relation "believing" has to that source and to our knowledge of it—that we have not yet achieved. What is incomprehensible is, in short, not mysterious; it is, rather, a matter of ordinary credal indifference.[9]

That Freud thought of the source of suffering as patently beyond belief accounts, I think, for the tone of deflationary coldness that many readers find so distasteful in his writings. In his theory of dream interpretation, for instance, he argues that the latent dream-thoughts, those thoughts we devote our days and nights to not believing, are straightforwardly expressible in simple language. The motive wish of even the most beautiful and intricate dreams is something simple like "I want my mother's love" or "I want my father's approval." There's nothing cognitively complicated in those thoughts, nothing that places special demands on knowledge. The source of our suffering, in other words, is not a complicated desire but rather a wishful, and therefore complicated, relation to a simple desire. From this perspective, our secrets turn out to be strangely open, even if not to us ourselves. This seems cold, of course, because for the dreamer the ungrasped thought is certainly lived as dark and dangerous and hard to hold in consciousness; for the psychoanalyst to regard it as simple, even innocuous, thus strikes us as dismissive of the dreamer's experience. And although there is undeniably something cold in that attitude, it is a coldness that is prerequisite to the analytic disclosure of how the dreamer suffers. This is why Freud insists that the subject matter of dream interpretation is not the latent dream-thoughts but rather what he calls the dream-work, that species of mental labor by means of which the dreamer works out her relations to her latent dream-thoughts. In a 1925 footnote to the chapter on secondary revision in the dream-work, Freud testily reminds his followers that latency is not secrecy but, rather, nonbelieved, that is, uncredited, patency:

> I used at one time to find it extraordinarily difficult to accustom readers to the distinction between the manifest content of dreams and the latent dream-thoughts. Again and again arguments and objections would be brought up based upon some uninterpreted dream in the form in which it had been retained in the memory, and the need to interpret it would be ignored. But now that analysts at least have become reconciled to replacing the manifest dream by the meaning revealed by its interpretation, many of them have become guilty of falling into another confusion which they cling to with equal obstinacy. They seek to find the essence of dreams in their latent content and in so doing overlook the distinction between the latent dream-thoughts and the dream-work. At bottom, dreams are nothing other than a particular *form* of thinking, made possible by the conditions of the state of sleep. It is the ***dream-work*** which creates that form, and it alone is the essence of dreaming—the explanation of its peculiar nature.[10]

So described, the dream is the measure not of the real darkness and danger of the latent dream-thoughts but rather of how even the simplest thoughts, when kept in the condition of latency, become dangerous wishes for the dreamer. In other words, the dream is the form in which is encoded and expressed the dreamer's fear of and compliance with the authoritative demand not to believe her dream-thoughts. In this light, the belief that the wish must be complex and fearsome is, far from being part of the truth about the dream-thought, part and parcel of submission to authority. That our secret thoughts are incomprehensible because we suffer from them is, in other words, a symptomatic judgment in which is secreted our complex relation to the authority that demands of us that we discredit those thoughts. The apparent coldness of his refusal to share the perspective of the patient's suffering notwithstanding, for the analyst to accept the mystery of the dream-thoughts would be to strike an alliance with the superegoic psychical force that banishes those thoughts to the realm of the incredible. This does not mean, of course, that the analyst in fact knows the thoughts in question, for those thoughts may be very well protected within the maze of defenses. Rather, the analyst does not endorse not-knowing as just desserts for those thoughts. To do that would be an exemplary failure to master the countertransference.[11]

Another way to put this point would be to say that what human beings suffer in their avoidance of the truth is not the truth but the avoidance. What we cannot believe, what we are obliged to make incomprehensible for ourselves, is the banal everydayness of the prohibition of our desires, and we do so in part by inflating the transgressiveness of the desires themselves. The measure of the grandeur of the gods to whom we bow down is the sacrifices they demand of us. In a secular voice: we are so staked to the disbelief of our desires that we create intricate cultures and elaborate institutions to maintain our disbelief. No mortal can keep a secret, Freud once mordantly wrote, by which he meant that the "secret" is so utterly out in the open that we are never done building new hiding places for it (*FCH* 77–78). This is also why dream interpretations, for Freud but for other psychoanalysts as well, usually break off, or simply tail off, long before the "secret" they contain is divulged. The point of the practice of interpretation is not to reveal the secret but to disclose the narrative that structures, and the personnel that staff, the demand that the dream-thoughts be guarded from view. All dream-thoughts are secrets open to someone, and it is the demand to defend against their openness that is the source of our suffering them as dreams.[12]

Now this view, that what is hidden to the dreamer or patient is, from some unattained point of view, really patent, sounds like a rank arrogation of authority by the analyst—and it would be, were the aim of psychoanalytic knowing grasping the reality of the source of suffering. But that, I am arguing, is not the psychoanalytic point at all. If it were, then psychotherapy really would be modern Socratism, which would be in utter contradiction with the central antirationalist discovery in the psychoanalytic consulting room that we do not cease suffering when we know the source of our suffering. The content of what we suffer is not hard to bring out. Indeed, we bring it out ourselves all the time! The loss we never make good on, yet which we never leave behind, is the very texture of our ordinary lives. What is hard to credit, which is to say what remains incomprehensible, is why we suffer it instead of believing it. What good (or god) is being served when we put smokescreens in front of ourselves? That is the great mystery that confronts the psychoanalyst, which is why the question psychoanalysis persistently asks in the face of the noncredibility of the obvious is: *Cui bono?* On whose behalf do we suffer it?

Although it is an oversimplification, I want to assert for now that both the aporetic nature of psychoanalytic inquiry as well as its indispensability arises on the site of this perplexity that knowing what we suffer is insufficient to generate conviction in our suffering it, and so also insufficient to address our suffering. Indeed, the ever-expanding catalog of both psychopathological and normal cultural forms that psychoanalysis is never done analyzing is composed of the disavowals and denials and isolations of what we already know, which is to say of the various forms of compliance with the compulsive demand to police our credal relations to the authority that exceeds the authority of knowledge. And this brings us back to an earlier point. It is not the thing not known by the patient that is the source of her suffering but rather the compulsion to comply with an incomprehensible authority, an authority that, because it ties incomprehensibility, as a psychical achievement, to unquestionability, can be described as obscene. The forms of compliance with this obscene authority keep the patent latent. But precisely because latency is simply patency discredited for the sufferer, psychoanalytic work cannot be simply a matter of diagnosis, which is to say it cannot be a matter of expert knowledge. ("Don't you see?" is a question that a psychoanalyst must never ask.) There is already expert knowledge written into the scene of suffering, but it is the all-knowing expertise jealously guarded by the father-imago who cannot be fled from. The leading question for psychoanalysis is not, therefore, "What do you keep hidden?"—that is the persecutory question of the internalized authority—but rather, "What are you powerless to keep for yourself?" The insight that the imago of authoritative expertise embodies an exercise of naked power forces psychoanalysis to turn the credo of Enlightenment inside out; rubbed against its grain, the imperative "Dare to become wise!" turns out to be hiding the father's threat, "Become wise? I dare you!" However much the psychoanalyst knows, it is never enough, never as much as is already known by the sufferer who is committed to protecting the authority for whom nothing is incomprehensible and so to whom she is bound. The practice of psychoanalysis is shaped, in short, by the insight that the limit of knowledge is a material limit, an embodied force, that is written into the texture of everyday suffering.

Now, this material limitation of psychoanalytic expertise clearly delivers to psychoanalytic metapsychology a sharp challenge, for what it implies is that the source of suffering cannot be identical with what it is known to be. Looked at from a certain distance, this is really just a logical point. To know the source of suffering as an origin is to know it as having happened at a point in the developmental past of the sufferer. That knowledge, however, cannot be perspicacious since it is in the nature of the suffering in question that the event has not ceased to happen. In other words, the suffering is the persistence of the (putative) past, which is to say it is, paradoxically, the past's not being over. More precisely: the suffering is the continuing force of the past in the present, the past's power to persist in and as fresh experience. The past is patently all over the place. Thus, any knowledge of the past as a discrete and discernible "it," as a quarantine for events, inappropriately privileges the power of knowledge to distinguish source from consequence, origin from aftermath. We can push this point a step further by noting that the freezing of the source of suffering at a distinct moment in the past is a form of defensive isolation. "This happened then"—the basic logical form of historical judgment—can be heard as the fulfillment of a wish to open an ordered temporal gap between the source and the suffering: "If only the past really were a foreign country!" What psychoanalytic metapsychology must grapple with, then, is the paradox that to the extent to which the source of suffering can be located in a developmental history, that history must be understood against its grain as a perpetually failed departure from its putative origin, as a developmental history that preserves the copresence of origin and aftermath. The origin keeps happening, and this must disable the prideful thought that to know the origin of suffering is to know it as the wellspring of suffering. A gap opens at the heart of psychoanalytic knowing—even, we might say, as the heart of psychoanalytic knowing.

Now, despite the tone of urgency with which I have just laid out the limitations of—the limitations internal to—the cognitive power of psychoanalytic metapsychology, it also needs to be said that all I have touched on so far is the good news. The copresence of past and present, the ability of the present to hold, and to hold on to, the undead past with which it is nonetheless in tension, is the mark of normal suffering. It is central to psychoanalysis, or at least to its Freudian and Kleinian variants, that suffering loss is constitutive of psychical life and that the capacity to sustain that loss through love and work is the measure of psychical health. Not transcending or redeeming loss, not ceasing to suffer, but sustaining loss, continuing to suffer creatively, is the mark of healthy individuation. Psychoanalytic metapsychology is, in this respect, not a theory of psychopathology but of normality. To see loss and suffering latently written across every text of culture and personality, and so to see the work of reparation as ongoing, is an affirmation of those texts, for it is the surest sign that the society that generates those texts is making possible the elaborations of loss and recovery that constitute nonorthopedic individualization. To say that loss and suffering are sustained in culture and personality is also to say that culture and personality are not the transcendence of need but, rather, need's orientational persistence. In sustaining culture and personality we deposit our suffering in the form of an incomprehensible demand, by which I mean a demand that calls for ever more elaborate and refined forms of acknowledgment. To the extent that the aim of psychoanalytic therapy is to renew our capacity to press acknowledgment's suit even at the "highest" levels of the cultural and cognitive practice, we can say that the aim of psychoanalytic metapsychology must be to make reflectively available the unending need to do combat with the superegoistic prohibition of mournful life. In sum, being able to think of loss and suffering as normal, and so being able to regard them in their inscrutable ordinariness as the basic textural elements of everyday life, is the needful challenge that psychoanalytic metapsychology in its self-critical specificity is shaped to meet. In that sense, my discussion of the patency of the unbelievable has been an excursus through territory that has been conquered (albeit without guarantee that it will remain so). That is what I meant by calling this the good news.

Traumatic Suffering

The conclusion of my argument up to this point is that incomprehensible loss is not the opposite of normal psychical life. It is, rather, the condition of its possibility. The core of our being (*Der Kern unseres Wesen*) is the losses we never make good on but are never done handling.[13] Now, I intend to devote the remainder of this chapter to exploring a discomfiting corollary of this conclusion: incomprehensible loss, in the sense I have discussed here, is neither identical with nor a condition for specifically traumatic suffering. Indeed, incomprehensible loss is not even a characteristic feature of traumatic suffering. I will return shortly to the significance of the erroneous identification of trauma with loss, but to get on the table as openly as possible my point about what seems specifically incomprehensible about trauma, and so how it works its way into metapsychological reflection, let me cite some passages from one of the most humane and theoretically sophisticated books on psychoanalytic traumatology, Stephen Prior's *Object Relations in Severe Trauma: Psychotherapy of the Sexually Abused Child*. "What happens to the psyche when it is overwhelmed," Prior writes, "is what theories of trauma need to explain."[14] An overwhelmed psyche, he goes on to explain, is one that has "internalized object relations that are intolerable, inconsistent, unstable, and unlivable" (Prior 104). The sufferer of trauma cannot live with the psychical introjects he is nonetheless fated to bear. (I will use the masculine pronoun here because Prior worked with sexually abused boys.) By "unlivable," which is a key analytic category for him, Prior aims to capture two connected features that are characteristic of traumatic suffering: (1) for the survivor of trauma, it is difficult to the point of impossibility to form new interpersonal and object relations because all new relations amount to nothing but the reliving—which prevents living—of what Prior calls the "overt" trauma, and (2) the repetition of the overt trauma is not just undergone by the sufferer but, even more direly, anticipated; new interpersonal and object relations are defensively avoided in fearful anticipation of reliving the trauma, which is to say that one of the debilitating symptoms of trauma is annihilation anxiety. Anxiety about annihilation is, we might say, not so much fear of this or that particular future but rather fear that there will be a future at all.

And not because of any uncertainty about the future. To the contrary: because the survivor of trauma is never done repeating the psychical contents introjected in and as his victimization, he cannot be said to have a future, in the sense of fresh experiences, but rather only a past that recurs in the same form over and again. When all futures are annihilating, it is certainty that terrifies. We might even say that traumatic suffering is the guarantee of a future that will be populated by the same old devils. Either way, the point remains the same: the traumatic overwhelming of the psyche can be understood as the fearsome repetition of the same, what Prior calls relentless reliving. Traumatic suffering, then, is characterized by the incessant return not of something incomprehensible, but of something immediately familiar. Far from being the same as the unbelievable loss that constitutes psyche, trauma is its opposite: in the aftermath of the traumatic blow, the sufferer has seen it all before, undergoes the necessity of his suffering, and rests confident that he will see it all again. Trauma is the absolute assurance that the past, present, and future cannot be mediated. It is the certainty of no experience.

Looked at this way, traumata must be understood not to disturb normal development but rather to disable normal development. The victimization is relentlessly relived, which is to say it recurs ahead of any possible developmental metabolization of it. Nothing is ever lost for the traumatic sufferer, nothing is ever recoded mournfully. Prior captures the specifically traumatic persistence of injury, the resilient impeding of development in which the abuser's power resides, by dubbing such traumas "overt." I find this a useful concept because it reminds us of how trauma protrudes from the field of experience, how it fails ever to secrete itself within fresh experiences that take their direction from the demand to mediate the past that conditions them. Overt trauma, in Prior's dialectically precise formulation, thus has "no developmental necessity." It opens no paths of connection, communication, or translation between incommensurable temporal registers. To the contrary: overt trauma is the fire that burns all bridges.

The overtness of trauma makes special demands on psychotherapy, to which I shall return in closing. However, it makes demands on metapsychology also, since in the destruction of developmental necessity the psychoanalyst witnesses not patently incomprehensible demand, which is the energy of development, but, rather, blatant force. Nothing is more open than pain in the moment of feeling it, nothing less demands a credal reorientation on the part of the witness, and thus nothing is more disabling to the drive for conceptualization. In an effort to express this theoretical disability, traumata are often referred to as "real," to distinguish them from the disturbances of phantasy that can be thought of as drive-derivatives. Prior adverts to this concept when he insists against Otto Kernberg and Fred Pine that the introjects of overt trauma are different in kind from those of normal developmental disturbances. Traumatic suffering is not the symptom of a constitutional inability to metabolize the residues of normal development. It is, instead, the appearance of something real in the place where phantasy ought to be, something real in the place where mindful activity ought to be. But "reality," as we know from Kant, is not a proper predicate, which is to say, it is not really a concept. The traumatic pain neither has a psychical place nor demands one. What it wants is not to be believed, but merely to be seen.

This all seems to me immensely illuminating, which is why it is so surprising that Prior also says:

> Freudian theory and its variants are all theories of trauma: the oedipal crisis, separation-individuation, and failures of self-object availability all present the psyche with inevitable yet overwhelming disruption that requires significant reorganization of the psyche to a "higher" level but also to a level where neurotic compromise is virtually inevitable. (Prior 119)

This cannot be right. The distinction Prior draws between overt trauma and other disturbances is that overt trauma has no developmental necessity, from which it follows that the oedipal crisis, arguably the paradigm of developmental necessity, is not traumatic. Admittedly, Prior's claim here is not that oedipalization, separation, and so on are overt traumas; he posits two classes of trauma, developmentally necessary trauma and overt trauma that destroys the necessity of development. But then, motivated by my earlier line of argument, we need to ask why we should think of "Freudian theory and its variants" as sweeping theories of trauma at all. Why not instead distinguish the losses that condition the possibility of development from the violence that impedes development? Why not distinguish disturbance from trauma or, if we prefer, distinguish radically between two forms of disturbance, one of which unsettles and so leads mind on toward its futures and one of which mortifies and so leads nowhere? Why not, in other words, think of psychoanalytic theory as an account of the contingent conditions necessary for a nontraumatized life? Doing so would enable us to avoid falling, as does Prior, from splendid premises into apocalyptic bathos.

> In the history of psychodynamic theory there is a powerful precedent of looking first at the more disturbed cases and then seeing what they reveal about the less disturbed, and then about all of us and human nature. Part of what is still shocking and difficult for people to accept about Freud and Freudian theory is that it is not a theory of psychopathology but a theory of normalcy. We are all disturbed. (Prior 14)

Up to this point, Prior is offering a splendid summary of the psychoanalytic idea that disturbance and development normally go hand-in-hand. However, he then turns in a radically different direction.

> Freud's theory is actually a theory of trauma, a fact obscured these days. The developmentally normal events of childhood—our prolonged dependence on our parents, the vicissitudes of our natural wishes, and the recalcitrance of reality—truly overwhelm every one of us and require those defenses and distortions we call neurosis. . . . The conflicts and forces that beset the overtly disturbed beset us all. (Prior 14)

The theoretical dilemma here is plain to see. Given the crucial specificity of Prior's notion of trauma as overwhelming interpersonal and object relations that are relentlessly relived and that persistently generate annihilation anxiety, it cannot be right to say that the developmentally normal events of childhood overwhelm us like that. Rather, those events shape us as those who are never finished developing. Loss happens to people, whereas traumatic violence, by contrast, eviscerates the prospect of any development of a psychical structure that might measure it. If we listen closely, we can hear Prior sensing that he has gotten caught in an overeager amalgamation of trauma and loss. Notice that he says of the "developmentally normal events of childhood" that they "truly" overwhelm us; that emphatic "truly"—"truly" as opposed to "apparently"?—betrays his anxiety about whether they really overwhelm us. Perhaps, Prior is thinking, they merely threaten to overwhelm us, and in that way give us orientation, albeit an orientation that, in the face of overt trauma, inevitably feels radically insufficient. In other words, in the face of traumatic suffering we are overwhelmed not by normal disturbances (which, as I have been arguing, do not truly overwhelm us) but rather by a different pain, an overt one, that cannot be mediated, cannot be measured, by everyday loss. Perhaps, then, we are all overwhelmed not by our own normal developmental disturbances but by the overt pain of the traumatized other. Traumatic pain is the demand that even proper orientation cannot meet. We are not, therefore, contrary to Prior, normally traumatized. Quite to the contrary: the mark of trauma is its palpable yet immeasurable distance from normality. The effort to assimilate normality to trauma, to make trauma the normal state of affairs, might seem to be an effort to honor trauma by bringing it under the concept of disturbance. In reality, however, it fails to acknowledge the blatant incomprehensibility that trauma makes all too vivid. The normalization of trauma by means of the reconceptualization of normal disturbance as trauma is a metapsychological evasion of the unmetabolized excess of trauma itself.

Allow me to put this point another way, even at the risk of overstatement. In its compulsive repetition of the moment of utterly vulnerable victimization, trauma is the incessance of injury. It is not the incessant return of the past, but rather the persistence of the thing itself. The thing itself is, of course, no "thing," no particular event brought to bay by means of predicative cognition, but rather, in its persistence, a broken-off piece of suffering. Trauma is a pattern of disorder. The vulnerability is perpetual and the work of metabolizing never commences. Trauma thus breaks the link discovered by psychoanalysis between the suffering of loss and development. Trauma, in this light, is not the disturbance at the heart of development but the violence that prevents the development of mediations of disturbance from starting. And it is because nothing ever gets lost in trauma that trauma represents the gravest challenge to psychoanalytic metapsychology. The authority of the psychoanalyst is the self-critical insight into the developmental significance of, precisely, conceptualization and its limits, so the thing of trauma, the thing it bears, is a direct challenge, an insult, even, to psychoanalytic critique. Freud pointed toward the problem when he observed of the death drive—the compulsion to return to the inorganic state that, while arguably at work in all living beings, seems to show itself in traumatic neurosis—that it never shows itself except as the destruction of all means of showing.

> If the self-preservative instincts too are of a libidinal nature, are there perhaps no other instincts whatever but the libidinal ones? At all events there are none other visible. . . . The difficulty remains that psychoanalysis has not enabled us hitherto to point to any [ego-]instincts other than the libidinal ones. That, however, is no reason for our falling in with the conclusion that no others in fact exist.[15]

The embarrassment here is patent and scrupulous. The sum total of experience offers no example of the concept "death drive," which means that empirical inquiry and scientific demonstration are stalled by experience. Yet to infer the absence of cause from the absence of example is to fall into the dogmatic trap of demanding an appearance of a force the nature of which is its implacable hostility to the normative demands of appearance. Thus, to honor the cause one has no option but to speculate. Only the unwinding of the claims of knowledge, the display alongside of cognitive achievement of the moment of cognitive weakness, can take the measure of the cause. Speculation, in other words, is the cognitive nonmetabolization of the need to think. Trauma, we must conclude then, gives rise to no development, not even within metapsychology. All one can do in the face of trauma is speculate, which is to say: all one can do is say, "I see." Not "I understand" (no conceptualization) or "I can't imagine" (no imagination) but rather only "I see" (speculation). Speculation is thought unbound, a cognitive undergoing of an experience that cannot find a place in the received patterns of conceptual knowing; speculation is the demand to know which plays itself out within a metapsychology that has registered the blow of the overtness of what it cannot but acknowledge; speculation is the mark of unmetabolized history. The authority of the analyst is demolished in this moment—"the reader will consider or dismiss [the speculation] according to his individual predilection," Freud says (*BPP* 24). In light of this, the urge to rebuild that authority by means of a metapsychological appropriation of the violence of trauma as a condition identical to the disturbances of normal developmental disturbance, as the same sort of problem as the one that metapsychology has already mastered, must be understood as a metapsychological denial of the violence in question. To assimilate trauma to normal disturbance valorizes violence, makes it over into the condition of development, and so blunts the demand to see traumatic suffering; to reconceive the disturbances of normal development as violence diminishes the value of the nontraumatic losses out of which is woven exactly that texture of everyday mindedness which remains our only measure of traumatic suffering. Trauma is cognitively infectious: it leads us mimetically to undermine the resources with which we might handle it.[16]

This entire set of claims is extraordinarily hard to get clear, so permit me to try it one other way. The remaking of all loss in the image of trauma is the first moment in the attempted restoration of the conceptual authority of psychoanalytic metapsychology, which authority has been confronted by an overtness that challenges the possibility of such authority. This sounds paradoxical, to be sure, since the effort to make trauma central to psychical development is undertaken in a humane effort, beautifully exemplified by Prior, to expand the Freudian insight that normality and disturbance are mutually conditioning rather than mutually exclusive. It is an effort to make room for what so powerfully challenges psychoanalytic conceptualization. Nonetheless, the conceptual work through which this effort to honor the traumatic is undertaken has, I am suggesting, the opposite consequence. It makes psychical normality so expansive that nothing can impede it, nothing can fall outside of its power. By means of a narcissistic inflation that philosophers are all too familiar with—it is the same mythological work done by idealism—the conceptual normalization of trauma strips psychical health of its insuperable contingency on circumstances that cannot be guaranteed a priori. This, I propose, is a manner of striving to bind the wound that trauma inflicts on metapsychology by recovering on the metapsychological level what remains unthinkable. But then it is a premature binding; the metapsychological assimilation of trauma to normal disturbance is the cognitive version of the motto with which our culture scorns all unsustainable pain: "Let the healing begin." Now, I do not deny that such idealization is a measure of the extremity and enormity of the wound (idealism in philosophy, insofar as it restores in the ideal order the consistent and reliable horizons of meaning disrupted by ineluctable contingency, is likewise the consequence of an encounter with persistent loss). I wish to emphasize, however, that the normalization of trauma is also a mimesis unto death, in the sense that it ceases to demand of psychical normality and from within psychical normality that it grapple with the demands of what remains other to it. The healthy psychical work of sustaining loss is measured as inadequate to and by the pain that never disappears, by the suffering that is never lost, but metapsychological inquiry, when it remakes normal disturbance as itself traumatic, encourages normal psychical functioning to stop seeing what is extreme for it. And this, I want to say, is a sort of cognitive disability. It defensively takes trauma inside the fragile and contingent achievements of psyche despite the overtness, the insurmountability, of the suffering. The universalization of trauma is, in that way, a denial of what in trauma is traumatic.

A Late Adventure of the Feelings

To identify trauma and loss is a double insult. First, it is an insult to normal psychical structure, because it makes normality depend on a capacity to measure the immeasurable, which is a sure sign of superego passionately out of control. Second, it is an insult to traumatic suffering, because it greets it as something familiar, as something we have seen before. And maybe—who knows?—we have seen it before. But in that specific imprecision of judgment resides the crucial and disturbing difference between trauma and normal disturbance: the overtness of trauma, and the cascade of failed mediations that tumble before it, makes traumatic suffering incomprehensible not because it is hard to credit but rather because it is impossible to credit. It is beyond belief, by which I mean to say, paradoxically, that it is all too believable. What we cannot swallow we cannot help but confront, and what we cannot help but confront, what we cannot rationally reconstruct but still cannot repudiate (or, perhaps, can only repudiate), is the greatest challenge to acknowledgment's power to repair moral and social injury. Trauma unfolds nothing—and therefore does not aim for acknowledgment. What is incomprehensible about trauma, then, might be put this way: although it remains ungrasped, there is nothing to be learned from it. Trauma gives rise to no concept or, if I may, "trauma" is not a concept. It is not even right to say that trauma challenges conceptualization, as if somehow the proper response to it is to think better. In the register of metapsychology, the fact of trauma is that we will never be able to think well enough. In the realm of ethics and politics, that means that we are always too late in the face of trauma. Trauma cannot be thought; it can only be halted. Not being able to acknowledge, I have argued, is the strength of the healthy psyche that probes for creative expressions of the excess stimulation of the persistently lost, so the ethico-political challenge of trauma amounts to this: we have not yet learned, even though we have felt the demand to learn, how not to acknowledge irreparable injury.

Let me close, then, with an opening. We are always too late when it comes to traumatic suffering, for at the heart of annihilation anxiety is the certainty that there is no more time in which to overcome the injury that should have annihilated the sufferer. The only question that remains before us is the question of belatedness or, to use Thomas Mann's idea, the question of a late adventure of the feelings. In the beginning of *Death in Venice*, Gustav von Aschenbach finds himself moved to travel, despite there being nothing of value he can imagine travel bringing him to. He is in a late phase in life in which he wants to finish his work in the time remaining, in order to work off the accumulated experiences of a life lived like a clenched fist. As his ship arrives in Venice, Aschenbach "ask[s] his own sober, weary heart if a new enthusiasm, a new preoccupation, some late adventure of the feelings, could still be in store for the idle traveler."[17] The late adventure of the feelings carries with it the desolation it cannot overcome (hence, Mann's peculiar figure of the "idle traveler") and so is a path to death. Indeed, Aschenbach takes up his death as fate when it becomes palpable to him that no path leads anywhere else. Still, that leaves open the question of adventure along the way: disenchanted adventure, adventure that does not redeem a life but that acknowledges its inevitable encounter with the same. The question that opens here is, then, that of a good death.

The conditions of such an adventure are, of course, strictly unimaginable, since they arise only when death ceases to be the aim of life—when annihilation anxiety ceases to rule the world. As Prior closes his book on the omnipresence of annihilation with a brief polemic, prosaically but precisely entitled "Counterindications," he asks what the conditions are under which trauma may be, not cured, but treated, not overcome but made livable. There are two nonnegotiable conditions he lays out. First, "the child must feel fundamentally supported by parents or the adults in the parental role." Given the abuse of Prior's boy patients by adults on whom they were dependent, this can mean only that the child must first be granted a fundamentally new adult world. Second, "the therapist must also feel supported and contained by the institutional setting of the therapy. He or she must feel safe with the child and with the primitive issues evoked by the treatment" (Prior 172–73). The therapist and the bearer of the wound must both have a world in which the incomprehensibility of the issues, their primitiveness, can be experienced within a new institution. New adults, new institutions: a new world, and a late one. Sanctuary from history, then, and not normalization, is the demand of traumatic suffering. And for that we must leave behind metapsychology and make our way toward late adventures of love and politics.

Notes

1. I owe thanks to many critics for helping me understand the purpose and structure of my arguments in this chapter, but to none more than Bettina Bergo and Kristen Brown Golden. Bergo, Golden, and Eric Nelson organized the massively stimulating interdisciplinary conference "Trauma: Reflection on Experience and Its Other" at Millsaps College in April 2003 at which I first presented these ideas. Since then, they offered me rigorous criticisms that have improved this chapter beyond measure. The many remaining infelicities of substance and style are, of course, my responsibility alone.

2. In this sense, the troubled relation of psychoanalytic metapsychology to empirical psychology of the mind is an instance of the Kantian claim that the condition of possibility of a science of mind is that there cannot be a single, complete, self-consistent perspective on mind. If we replace "reason" with the "unconscious" in the following passage in which Kant takes a major step in constructing his compatibilist account of freedom and determinism, then the timelessness of the unconscious, and so its opening to a scene of mental causality different from natural causality, bounds across the century separating Kant from Freud.

> Man is himself an appearance. His will has an empirical character, which is the empirical cause of all his actions. There is no condition determining man in accordance with this character which is not contained in the series of natural effects, or which is not subject to their law—the law according to which there can be no empirically unconditioned causality of that which happens in time. Therefore no given action (since it can be perceived only as appearance) can begin absolutely of itself. But of pure reason we cannot say that the state wherein the will is determined is preceded and itself determined by some other state. For since reason is not itself an appearance, and is not subject to any conditions of sensibility, it follows that even as regards its causality there is in it no time-sequence, and that the dynamical law of nature, which determines succession in time according with rules, is not applicable to it.

See Immanuel Kant, *Critique of Pure Reason*, trans. Norman Kemp Smith (New York: St. Martin's Press, 1965), 476.

3. Stephen Mitchell and Margaret Black, *Freud and Beyond: A History of Modern Psychoanalytic Thought* (New York: Basic Books, 1995).

4. Sigmund Freud, "Fragment of an Analysis of a Case of Hysteria," in *The Standard Edition of the Complete Psychological Works of Sigmund Freud*, vol. 7, trans. James Strachey and Anna Freud (London: Hogarth Press and the Institute of Psycho-analysis, 1953), 122. Hereafter cited in the text as *FCH*.

5. Jonathan Lear, "Knowingness and Abandonment: An Oedipus for Our Time," in *Open Minded: Working Out the Logic of the Soul* (Cambridge, MA: Harvard University Press, 1998), 33–55.

6. Schafer has worked extensively on narrative, narrative theory, and psychoanalysis. *Retelling a Life* (New York: Basic Books, 1994) is, in my view, his most illuminating account, but also consult the chapters "Narration in the Psychoanalytic Dialogue" and "Action and Narration in Psychoanalysis" in his *The Analytic Attitude* (New York: Basic Books, 1993), 212–56. The lecture in which he doubted the authoritative expertise of the analyst was delivered at a conference at Columbia University on psychoanalysis and philosophy in 2000.

7. Adam Phillips, *Terrors and Experts* (Cambridge, MA: Harvard University Press, 1996), esp. 1–32. Lear canvasses similar territory in *Therapeutic Action: An Earnest Plea for Irony* (New York: Other Press, 2003).

8. Jacques Lacan, *The Four Fundamental Concepts of Psychoanalysis* (New York: Norton, 1978), 230–43.

9. By "credal indifference" I do not mean lack of credal interest. Establishing and maintaining credal indifference toward what we know requires devoted and active interest. The bearers of this interest carry familiar psychoanalytic names: disavowal, hostility, repudiation, and so on.

10. Sigmund Freud, *The Interpretation of Dreams*, in *The Standard Edition of the Complete Psychological Works of Sigmund Freud*, vol. 5, trans. James Strachey and Anna Freud (London: Hogarth Press and the Institute of Psycho-analysis, 1953), 506–7.

11. What appears from the analysand's point of view to be the analyst's refusal to embody the subject-supposed-to-know, even when it is obvious to everyone in the room just how bad the analysand's latent thoughts are, is also the source of one of the great analytic comedies of imprecation: the analyst's refusal to endorse superegoic condemnation also bars the analyst from embodying the subject-supposed-to-forgive.

12. To suffer the latent thoughts as dreams—or worse. While dreams have their roots in the same principles of psychical functioning as neuroses, the suffering of dreams and of neuroses should not be thought of as identical. In dreams we keep secret from others what is secret for ourselves, whereas neurotics are obliged, under command from a stern Goddess, to confess what they suffer. See Sigmund Freud, "Creative Writers and Day-dreaming," in *The Standard Edition of the Complete Psychological Works of Sigmund Freud*, vol. 9, trans. James Strachey and Anna Freud (London: Hogarth Press and the Institute of Psycho-analysis, 1953), 146.

13. Sigmund Freud, "An Outline of Psychoanalysis," in *The Standard Edition of the Complete Psychological Works of Sigmund Freud*, vol. 23, trans. James Strachey and Anna Freud (London: Hogarth Press and the Institute of Psycho-analysis, 1953), 197.

14. Stephen Prior, *Object Relations in Severe Trauma: Psychotherapy of the Sexually Abused Child* (Northvale, NJ: Jason Aronson, 1996), 13. Hereafter cited in the text as Prior.

15. Sigmund Freud, *Beyond the Pleasure Principle*, in *The Standard Edition of the Complete Psychological Works of Sigmund Freud*, vol. 18, trans. James Strachey and Anna Freud (London: Hogarth Press and the Institute of Psycho-analysis, 1953), 52-53. Hereafter cited in the text as *BPP*.

16. Bettina Bergo has challenged me to make my thoughts about the role of speculation in the nonmetabolization of trauma more concrete, especially in light of the philosophical ideal of speculation as, precisely, a higher-level cognitive reworking of what resists empirical conceptualization. Her request for further specification seems to me exactly right, for the very idea of conceiving the speculative discourses of philosophy, poetry, metapsychology, and so on, as cognitively significant at all depends on their cultivation of the capacity to bear the demand, issued by incomprehensible loss, to be known, which is to say the demand to be taken up by minded creatures capable of acknowledgment. All the same, I am not entirely confident how to satisfy Bergo's request. Speculation does honor to incomprehensible loss, in the sense that it opens a space of cognitive engagement past the horizon of inherited modes of empirical conceptualization. Perhaps it will always remain a matter of taste, but I suppose nonetheless that making good on the honor paid to incomprehensible loss cannot finally happen on the merely speculative level, and that at some moment what is lost must be seen to appear—massively mediated, to be sure—in the world, and so as capable of some sort of predicative cognition that actively, that is, empirically, interferes with inherited modes of empirical conceptualization. The truth of philosophical idealism (as of aesthetic idealism) is, I am saying, located in its deferral of the identification of predicative and speculative cognition; its falsity is located in its segregation of speculative cognition from the utopian promise of predicative cognition. But the point I am working toward here is, in a sense, at right angles to the problem of the relation between loss and speculation. For if it makes sense to say, as I have here, that "speculation is the cognitive nonmetabolization of the need to think," then I am leaving it open whether traumatic suffering is even a candidate for speculative cognition. My thought is that it is not, and that speculative knowing is not what traumatic suffering needs from minded creatures. I will suggest in my conclusion, however, that traumatic suffering nonetheless makes material demands on cognition, albeit demands that cognition alone cannot satisfy.

17. Thomas Mann, *Death in Venice and Seven Other Stories*, trans. H. T. Lowe-Porter (New York: Vintage International, 1989), 18.

3

Overcoming the Confusion of Loss and Trauma

The Need of Thinking Historically

Sara Beardsworth

Introduction

This chapter is on why it remains important to retain psychoanalytic thought at the center of our attempts to think through the source, nature, and significance of trauma. The focus here is less on developing a distinct conception of trauma than asking why the concept of trauma keeps turning up in the wrong place. What is at issue is the error, appearing more or less sporadically, that takes psychoanalytic thought to have rooted subjectivity in original trauma.[1] The claim here is that where and when this mistake does turn up it depends on a more general confusion of trauma with constitutive loss. In Freud, a trauma is an event in subjectivity that overwhelms the ego's defenses, and which is not itself experienced. The event is not traumatic but traumatogenic: the trauma appears only in belated symptoms, the sufferer remaining tied to a past that was never present and does not pass by. In brief, trauma in Freud is the breaking of experience. In contrast, constitutive loss is the loss of a primary object—paradigmatically the mother's body—that is necessary for individuation. A subject can only come into being on the basis of a lost object. The loss of the other is a vital condition of futurity. Nonetheless, to recent psychoanalytic experience, contemporary subjectivity in modern Western societies exhibits a general loss of loss, that is, psychic life shows evidence of a failure to undergo loss. This chapter argues that this loss of loss is connected to the appearance of the idea of original trauma, and that this can only be discovered if both loss and trauma are dealt with historically. I propose that the confusion between the two belongs to our inability to think through the fundamental nature of the loss of the other. This inability inhabits our modes of historical reflection, meaning, here, both child development theory, which explores subject formation (the maturation of the individual), and the analysis of social and cultural formation.

The historical reflection present in psychoanalysis is an attempt, beginning with Freud, to reveal the connections between subjective and social formation. Psychoanalysis is widely known as an investigation into the trials of separation. The infant must come to separate from radical dependence on another who is not yet an outside other. Selfhood comes into being in and through the gradated forming of relation to otherness. Since, for psychoanalysis, separation is therefore relation, the investigation into the trials of separation is equally an inquiry into the possibility of bonds with others and socialization. However, the connections between subjective and social formation are often missed in the reception of the key concepts that develop the psychoanalytic exploration of the trials of separation. The key concepts are primary narcissism and the Oedipus complex, two consequential stages in the maturation of the individual. Briefly, primary narcissism turns on the early dependence on the mother's body and presents the most primitive moments of ego formation that develop within the dual relationship of mother and child. The Oedipus complex appears later, reaching its peak between the ages of three and five, according to Freud. It has a triangular structure composed of the maternal apex as object of infantile affection (the incestuous object), the paternal apex as the prohibition on the infantile wish not to give up the attachment to the mother, and the infantile ego in formation together with conflictual impulses. Oedipus is Freud's figure of the difficulty and undesirability of separation when there is no turning back but the strength of the erotic wishes—affection for the mother—means that separation is established only in and through a struggle with a prohibitive paternal law.

When psychoanalysis discovers primary narcissism (Narcissus) and the struggle with paternal law (Oedipus) in the psychic life of adults, it proposes that they are not simply episodes in infantile life but are retained as fundamental structures of subjectivity. There is evidence for the view that psychoanalysis takes them to be universal and transhistorical structures. Laplanche and Pontalis underline the foundational character that the Oedipus complex had for Freud, since it is not a matter of actual parental agency but of the function of the proscriptive agency that impels separation.[2] Moreover, since this function is not attached to the empirical parental situation—and because the separating function is supported by an imaginary paternal threat, the threat of castration—the proscriptive agency is a myth with foundational properties. Freud's oedipal structure is rightly known as the psychoanalytic myth of origin. The foundational properties acquire universal cultural scope in the development of psychoanalytic theory when Lacan ties the oedipal structure to the achievement of the entrance into language or the symbolic order. Lacan presents the prohibitive function as the symbolic, paternal function whose internalization in the process of subject formation underpins the development of linguistic capacities, and so opens up social and symbolic being. That is, Lacan turned the paternal, prescriptive function into the foundation of culture itself. Later investigations into the narcissistic structure of subjectivity suggest, however, that the primacy of the triangular oedipal structure in separation is questionable. A deeper understanding of features that support ego formation within the earlier dual relationship of mother and child, known as primary narcissism, gives more weight to this logically and chronologically prior moment in the evolution of the individual. Nevertheless, whether the focus is on the narcissistic structure or the oedipal one, once it is thought that psychoanalysis presents these moments in subject formation as traumatic, the conclusion is drawn that psychoanalysis makes trauma originary for subjectivity, as though trauma were foundational for separation, and thereby bonds with others; as though it opened up futurity.

This chapter has three main objectives. The first is to draw out the resources that exist in psychoanalysis for comprehending the cultural determinations of these structures of subjectivity, revealing them to be the unconscious structures of modern subjects. The second is to show that Freud and Kristeva together offer not a conception of original trauma but genealogies of subjectivity. I first turn to Freud on Oedipus for the connection of the two modes of historical reflection, subjective and social, that appears in his analysis of the source of religion and religion's relation to its source. This analysis presents a genealogy of the Oedipus complex. I then turn to Kristeva's reassessment of Narcissus, the earlier moment of subjectivity, for her rethinking of the maternal role in subject formation. The argument is that if we attend to Freud's thought on the father and Kristeva's on the mother, we find that the psychoanalytic message is that what is originary for subjectivity is an immemorial exposure to absolute otherness, and that it is the loss of the other that is foundational for subjectivity. However, their respective thinking on the loss of the other, appearing in radically different contexts, in fact offers two distinct genealogies of subjectivity, and what is remarkable about them is that they cannot be unified. This chapter unfolds those two genealogies and underlines the impossibility of unifying them. Its third major objective is to show that psychoanalysis also opens up a genealogy of trauma itself. The conclusion to the chapter will draw out how and why the confusion of trauma and constitutive loss turns up in the historical reflection of modern thought.

Freud's Oedipus: The Trace of an Absent Religion

The discussion in this section of Freud's thought on the father has the following specific aims. I return, first, to the connection between religion and the structure of the psyche in *Totem and Taboo* (1912–13) in order to show how it illuminates the meaning of the Oedipus complex for Freud. What is crucial here is that the complex is understood, not only as the widely known conflict between desire and fear, but also as the site of ambivalent feelings toward the father: love and hatred. This ambivalence is both foundational for polytheistic religion, in the form of totemic systems, and central to the oedipal structure of the psyche, according to Freud. I then show that his later thought on religion, appearing in *Moses and Monotheism* (1939), unfolds the full significance of Freud's discovery of the oedipal structure of subjectivity. This is where the genealogy of the Oedipus complex turns up. We will then see that Kristeva's thought on the mother reveals the limitations of Freud's own historical reflection.

Freud's *Totem and Taboo* is an analysis of the formation of the sacred, that is, of the totemic system, and an interpretation of the two fundamental laws of totemic religion: the incest taboo and the taboo on murder. Let us briefly recall what is commonly taken to be the central moment in this thesis before considering the wider question of Freud's connection of religion and the psyche. The central moment is the oft-quoted passage on how the taboos come to be installed. For Freud, this is equally the manner in which the first forms of human society arise. The famous passage posits a primal horde in which a violent and jealous father kept all the females for himself and drove away his sons once they were grown. One day the brothers came together, killed and devoured the hated but admired father, and so made an end of the patriarchal horde.[3] The consequences of this act rest on the fate of the sons' ambivalent feelings toward the father. The sons' hatred is satisfied by the act of murder and their affection for the admired father reappeared in the shape of the remorse felt for the deed. "The dead father became stronger than the living one had been," in accordance with a psychological procedure of deferred obedience. The sons "revoked their deed by forbidding the killing of the totem animal, the substitute for their father; and they renounced its fruits by resigning their claim to the women" (*TT* 143). This, for Freud, is the immemorial past and hidden meaning of the two laws of the totemic system: incest taboo and the taboo on murder. Above all, since it is the vanquished (murdered, dead) father who is the source of paternal authority—that is, since it is deferred obedience to the father that Freud discerns behind the taboo on killing the totem animal—socialization begins with identification with the function of authority, whose meaning is opaque.[4] This is neither overt identification with the father nor express obedience to paternal law. Indeed, the figure of the father is abolished in the totemic system, whose social organization is that of the fraternal clan. Patriarchal society properly speaking arises once the dead father resumes his human shape in the form of a deity: not the One God but the conception of deity as such. The father is now present, if not recognized. Thus, first, paternal authority is at its height as a social authority in religion, and, second, polytheistic religions maintain the magical, taboo structure of authority.

How, then, does Freud connect religion and the structure of the psyche? At this point the relation between the two is simply an analogy on the grounds that the father complex is one source of religion and is found at the core of the psyche. What determines a social reality in the sacred and religion—paternal authority and emotional ambivalence in relation to the father—is the psychical reality that Freud discovers as the nucleus of neurosis, for the incest taboo and the taboo on murder turn up in the Oedipus complex. In other words, what has been a social reality is internalized and survives as a psychical reality, which is, as Freud underlines, asocial, neurotic (*TT* 74). In sum, both religious man and the structure of the psyche Freud discovers in modern man inhabit the taboo structure of morality and socialization. Nonetheless, this thesis does not help much overall in answering the question whether or not the Oedipus complex is universal and transhistorical. This is because, in proposing the analogy between religion and the psyche in *Totem and Taboo*, Freud didn't need to think through how religion itself is actually constituted: what the source of the authority of religion itself actually is. This task is taken on in the later work, *Moses and Monotheism*. Here Freud unfolds the core of the structure of traditional, paternal authority in a manner that explicates his discovery of the unconscious structure of modern subjects. *Moses and Monotheism* sets out what is required for a traditional authority to get established at all, and how the psychical reality of modern subjects is dominated by the myth of origin. It takes Freud's shocking interpretation of the formation and establishment of Hebraic monotheism to show what he is really saying about separation and socialization.

As we will see, Freud's revisions to the official account of the emergence of Hebraic monotheism appear to root it in an original trauma. Given the analogy between religion and the psyche established in *Totem and Taboo*, the message then appears to be that subjectivity is rooted in a primal trauma. Trauma becomes the origin of subjectivity. This is the mistaken impression that needs to be overcome by showing that, in Freud, the constitutive source of religion as a social reality is not trauma but loss. What then needs clarifying is the relationship between religion and the psyche, which is no longer left simply as an analogy but is given a genealogical treatment. Freud's *Moses and Monotheism* can be read as a thought on how a "beginning" in exposure to an absolute, immemorial otherness is negotiated rationally and historically. He focuses on religion because he finds that the structure of the psyche retains fundamentally the same elements as the religious negotiation of the source of the authority of religion.

Freud's interpretation of the social and historical establishment of the Mosaic doctrine is shocking because he both turns Moses from a Jew into an Egyptian and overturns the official doctrine that Hebraic monotheism begins in liberation from Egypt. On Freud's account, the "chosen people" are formed through the imposition of a foreign religious doctrine on the Semitic tribes scattered on the margins of Egypt. Moses was a great foreigner, an Egyptian, who reinstated the short-lived universal religion of one Pharaoh (Akhenaten). This moment of an Egyptian universal religion was quickly suppressed in that land, and buried under a revived polytheism. Freud's hypothesis is that Moses was a once-powerful official of the universal religion whose downfall led to his longing for its reascendancy. Moses "sought compensation for his losses."[5] When the Semitic tribes are elected to recover and fulfill this lost past, the chosen people are submitted to a truly monotheistic religion: a deity that is not only universal within the society but the only god—the One God. One advantage of Freud's interpretation is that it clears up the puzzle of the idea of the chosen people: the remarkable event that a people is neither tied to its deity from eternity nor chooses its deity but is, instead, chosen by it.[6] The Mosaic doctrine rests on the idea of election because it is the doctrine of a foreign God, one who disdains sacrifice and ceremonial and asks only for faith and a life in truth and justice. For Freud, the Mosaic doctrine is made up of ethical demands whose nucleus is a prohibitive law enforcing an abstraction from the senses. The God of Moses is an absolute power that cuts the faithful off from the pleasures returned to in the rites and practices of polytheism, ejecting them from all they have known, and leading them into, not liberation, but exile. On this view, "being chosen" is not a liberation from the power of the Pharaohs but a traumatogenic event.

Freud's concept of trauma is attentive to the temporal structure displayed in traumatic suffering. A trauma is an event that overwhelms the ego's defenses. We are given one form of egoic defense in the theory of the repression of instinctual impulses. Instinctual demands that are experienced as overstimulating because they threaten the stabilization of the ego undergo repression. They are excluded from the organization of the ego and relegated to the id, or "that," a term for the unconscious life of repressed impulses, and one that Freud deploys in order to capture how the internal pressure of the renounced impulses is felt by the subject to be coming from something alien. However, egoic defenses are also directed to the maintenance of an inside/outside border and protect against external phenomena that threaten to be overintrusive. The inside/outside border is necessary for there to be objects for a subject. The object in psychoanalysis is the correlate of the ego and without this fundamental correlation there can be no experience of objects. An event that overwhelms the ego's defenses cannot appear within the subject-object correlation. That is, the event does not appear for, or cannot be present to, the subject. It is not "framed by consciousness." Thus, trauma in Freud is an event that is not itself experienced. The event itself is marked by a lack of affect and the situation of danger is already past when the trauma awakens. In sum, the event is traumatogenic, which is to say, the trauma appears only in its repercussions: symptoms that are belated with respect to what triggers them. Freud finds evidence for this temporal structure of trauma within official doctrine itself when the latter acknowledges that Moses's followers return to polytheistic rites in their condition of exile. Episodes such as that of the golden calf reveal the reaction against the severity of the moral doctrine, as the Semites joyfully return to the pleasures of sacrifice and idolatry. The reassertion of renounced aggression toward a powerful father appears in the pleasure-seeking that provides satisfaction for the repressed erotic impulses ruled out by his will. However, the core of renounced aggression is the repressed murderous impulse felt toward the overpowerful father, and Freud's story of Moses proceeds on the hypothesis that the people whom Moses chose for his followers acted on the impulse. The great foreigner was murdered by his God's chosen people. The murder, the doctrine, and the real conditions of exile are repressed.[7] Freud's story of Moses therefore presents a prehistory of Hebraic monotheism in which murder and repression are the fate of trauma.

The question is, then, what conditions make it possible for the Mosaic doctrine to reassert itself as a great moral and social structure? Freud struggles to link his story of Moses with the thesis of *Totem and Taboo*, in which the phenomenon of emotional ambivalence toward the father of the primal horde allows for the reassertion of paternal authority in the wake of the vanquishing of the father. The vicissitudes of love and hatred found a dialectic of prohibition and transgression that accounts for the formation, deformation, and reformation of polytheistic religion. Emotional ambivalence is foundational for polytheism because, while the hatred for the father is carried through in the murderous deed, affection for the father resurfaces in its aftermath in the form of remorse. Remorse reveals the sons' identification with the vanquished father. In other words, the oedipal dialectic known from *Totem and Taboo* unfolds from the twofold nature of the relationship to the (loved and hated) father, whose correlate in the structure of the psyche is the ego ideal, on the one hand, and the punitive superego, on the other. In *Moses and Monotheism*, in contrast, no such dialectic is present, or seems possible, for there is no prior relationship that would allow for the factor of emotional ambivalence. In essence, the phenomenon of identification with the father is lacking to the chosen people. This father is not theirs. It is therefore hard to see how the reassertion of the Mosaic doctrine is possible. Freud himself speculates on the presence of phylogenetic factors that would allow him to place emotional ambivalence—and so the necessary phenomenon of identification—at the moment of election by the foreign God and its immediate fate. He proposes that the appearance of the Mosaic doctrine and the murder of the man Moses activate the unconscious memory of the age-old and long-forgotten relationship toward the primeval father, with all of its emotional ambivalence. That is, the identification of the chosen people with the Mosaic God is accounted for by a return of the repressed.

Clearly this solution to Freud's problem gives the myth of origin he discovers in the structure of the psyche a universal and transhistorical status. It does nothing to illuminate the mythical quality of the law of the psyche. However, there are other aspects of Freud's story of Moses that give the return of the repressed a specific role in the establishment of monotheism as distinct from polytheism. We will take the complete absence of the phenomenon of remorse in Freud's story of Moses, and his assertion that the Mosaic religion influenced the Jewish people only once it had become a tradition, to suggest that he must look elsewhere for the possibility of the reassertion of the Mosaic doctrine and its triumph as a moral and social structure. In finding this possibility elsewhere in Freud's account, I will show, first, that the phenomenon of identification can and must turn up at a moment later than the event of election and the murderous deed. Second, the discovery of what the source of the authority of religion actually is will remove the false impression that Freud makes trauma the source of religion. Third, I will show that Freud's analysis of the cultural determinants of the oedipal structure of the psyche in *Moses and Monotheism* does in fact deliver a genealogy of the Oedipus complex.

It is notable that Freud's later thought on the function of paternal law in the evolution of the individual bears a stricter resemblance to the story of Moses than that of the fraternal clan. The intolerable "tones of a foreign God"—as Kristeva aptly puts it[8]—find their echo in the paternal prohibition of the oedipal father; the rupture from the pleasures of polytheism in the repression of erotic wishes; and the condition of exile in the condition of separateness or individuation as an ejection from the infantile home. It is Freud's attention to the fate of the internalized paternal law—that is, to the formation of the superego—that reveals the lesser import of the *Totem and Taboo* legend in his later thought. In *Civilization and its Discontents* (1930), Freud explores the significance of the superego as a tyrannical agency formed out of a split-off component of the ego that combines three central features: the internalized paternal law, the aggression feared from the father, and the renounced aggression felt toward him. In its treatment of the ego the superego makes no distinction between the murderous impulses whose pressure comes from the id and the commitment of the deed, which would be an act of the ego. What Freud is drawing out is the extent to which the oedipal structure is a guilt structure of subjectivity. The oedipal structure is now known as one that ties the subject to an unconscious sense of guilt, as though the factor of affection for the father had disappeared. Above all, what is missing is the image of the loved father or the ideal ego—identification with the father—as distinct from the punitive superego. We will return later to this question of the disappearance of the loved father in our discussion of Kristeva's thought. What is apparent at this point is that Freud's account of the foundation of the authority of monotheistic religion now seems to be more pertinent for the connection between religion and the psyche than his earlier account of the development of polytheism.

In order to comprehend this we must return to Freud's version of the Exodus and the establishment of Hebraic monotheism, for we have not yet fully grasped how the authority of religion is actually accounted for. Thus far we have understood the protracted prehistory of Hebraic monotheism on the basis of election by a foreign God, followed by the murder of Moses, and repression. The establishment of Hebraic monotheism as a moral and social structure is therefore, as Freud underlines, a return of the repressed. Yet if we abandon the notion that the latter comes with emotional ambivalence toward the father, the foundational properties of ambivalence, known from *Totem and Taboo*, are missing. Some other element must be present in the prehistory of monotheism that makes up for this lack. It is tempting to find it within the traumatogenic event of being chosen or, better, the temporal structure of trauma itself. However, the latter does not allow the traumatogenic event to be a founding event. With Freud, traumatic suffering is caught in a present that is nonfutural because the sufferer cannot take up a relation to what has occurred. Traumatic suffering is structured by the persistence and dominion of a past within the present, turning it into an endless present.[9] If, in *Moses and Monotheism*, murder and repression are the fate of trauma, there must be some element that illuminates the possibility of the reassertion of the doctrine—the return of the repressed—but does not involve a resurfacing of the traumatogenic features of its first appearance and so a recycling of that fate.

This element is latency. A crucial factor in the establishment of Hebraic monotheism as a moral and social structure is the long-protracted nature of its prehistory: a latency period in which the Mosaic doctrine persists only in a form that is "split off" from what is recognized and lived—persisting perhaps in oral transmission, but only slowly coming to reassert itself as the religion of the Jews (*MM* 68). In our effort to comprehend what enables the reassertion of Mosaic law as a social reality we can take as a guide the fact that the return of the repressed is never the recovery of "what was"—some past event or content. Freud underlines the factor of distortion in the return of the repressed. For him, a distortion both secretes and indicates abnegated material. Yet he also stresses the meaning of distortion as displacement or a kind of transformation (*MM* 43). We have noted the profound implications of this phenomenon as it appears in *Totem and Taboo*. The transformation of affection for the father into remorse felt for the murderous deed is foundational for the emergence of social structures in the fraternal clan. It is also what establishes the fundamental link between socialization and morality: the taboo structure of the totemic system. In *Moses and Monotheism*, in contrast, it is the factor of latency—Freud's repeated emphasis on the long drawn out development of Hebraic monotheism—that is crucial. The extended period in which the Mosaic doctrine persists in a split-off form carries out a transformation of the traumatogenic past—the intolerable commands of a foreign God—into the loss of the other: Moses, the great foreigner. This absolute loss is what enables the binding of past, present, and future in the religion of the Jews—a binding that trauma itself undoes. The possibility of the establishment of Hebraic monotheism as a moral and social structure therefore lies in the transformation of trauma into loss. This is what constitutes the authority of religion for Freud, but it is what religion does not know, what it forgets. In sum, what psychoanalysis thinks was the source of the power of religion was that, unbeknownst to itself, it was grounded in absolute loss. The loss of loss structures the obsessive quality that Freud finds in religion, a quality that he develops in *Moses and Monotheism* in a long reflection on the persisting presence within Hebraic monotheism of the great man: Moses the Jew.

We have followed through Freud's thesis on religion, which is to say, his analysis of the beginnings of the formation of Judeo-Christian culture.[10] It can now be seen that the whole import and possibility of Freud's connection of child development theory with social and cultural analysis turns on religion's rational and historical negotiation of the source of its authority—in other words, on the loss of loss. That is, Freud's genealogy of the Oedipus complex is to be found in the historical unfolding of the loss of loss. In the first place, religion takes the power of Mosaic law to bind and rule the chosen people to be based on the commands of religion itself. Traditional authority is the generational transmission of an inherited past. However, the authority of religion itself remains magical and unnegotiable because the rational and historical negotiation of religion with its source in the exposure to and loss of the other fails the loss. Traditional authority is won through the identification of the faithful with the commands that formulate truth and justice, an identification that turns up only given the latency period and its transformation of trauma into loss, but which is itself structured by the loss of loss and all that this implies for the identity of the Jews. Although Freud acknowledges the advance in culture represented by a religion in which emphasis is laid on ethical demands—the only "pure" monotheism in his view—he underlines that the ethical ideas cannot "disavow their origin from the sense of guilt felt on account of a repressed hostility to God" (*MM* 134). The sense of guilt is felt owing to the tie between the repressed hostility and the prohibitive law. That is, the identity of the chosen people upholds the prohibition on and repression of desire, and this is the failure of religion from a psychoanalytic standpoint: its fragility and tendency to collapse. In the second place, the relation of religion to its source is preserved in the secular aftermath of religion. Freud discovers the mode of its preservation when he encounters the unconscious oedipal structure of modern subjects. The oedipal structure of subjectivity that Freud discovered and maintained at the center of his therapeutic and theoretical endeavors is the discovery of the dominion of an infantile struggle: a struggle between desire and fear carried out in relation to paternal law despite the demise of paternal authority as a social reality. The elements of religion and religion's relation to its source are internalized. They become a psychical reality. In sum, the psyche in Freud is the trace of an absent religion: a past that has not passed by. The psychoanalytic claim that a myth of origin forms and dominates the psyche is therefore the claim that religion begat the psyche.

What is particularly fearsome in this situation is that what determines the failure of religion now persists in the unconscious structure of modern subjects in the most intransigent and unnegotiable form: unconscious fear of the superego. The core of the oedipal structure of subjectivity is a tension between the ego and the superego, the former measuring its own strivings against the latter's harsh demands. In other words, Freud's modern Oedipus occupies the guilt structure of subjectivity—what he called "a permanent internal unhappiness" in *Civilization and its Discontents.*[11] This reveals the persistent authority of the absent father, still suffered in his absence. Freud's remarks on ethics in the secular aftermath of religion are very clear:

> a part of its precepts are justified rationally by the necessity for delimiting the rights of society as against the individual, the rights of the individual as against society, and those of individuals as against one another. But what seems to us so grandiose about ethics, so mysterious and, in a mystical fashion, so self-evident, owes these characteristics to its connection with religion, its origin from the will of the father. (*MM* 122)

In other words, apart from the legal domain of the arrangement of rights, ethics in the secular aftermath of religion—or the essence of conscience, as Freud would say—is no more than superegoic prohibition: the taboo on incest and the taboo on murder. That is, the authority of morality remains nothing other than the taboo, a magical, and thereby unnegotiable structure of authority. It cannot be thought through. Freud's thought on the myth of origin is therefore, in its full scope, an analysis of a culture in which adequate selfhood and connection with others are undermined because it is a guilt culture. Put otherwise, he finds that the maturational processes typically available for subjectivity in Western cultures are inadequate. The murkiness and intransigence of this condition shows up in the confusion over what is and what is not historical, or culturally delimited, in the psychic structures. Yet the analysis of the formation of Judeo-Christian culture has shown that the elements of guilt culture lie in the rational and historical negotiation of the constitutive exposure to, and loss of, the other.

It is clear that historical reflection on tradition in the secular aftermath of religion has uncovered the traumatogenic past of traditional authority. Of course, what appears as chronologically first in a social and historical structure is not, for that reason, origin. This is something that psychoanalytic thought on repression, latency, and the return of the repressed gets at in a powerful way. Its historical thinking has offered a vital thought on the loss of the other as the source of religion, the loss of loss in religion's relation to its source, and how religion begat the psyche. This thought is failed, again, when it is assumed that psychoanalysis presents trauma as original for subjectivity. Even so, something remains opaque in the psychoanalytic account of Oedipus. It must be asked what conditions could lead a past social reality to be preserved as an exclusively psychical reality. Attention to this question is required if we are to fully approach the problem of the confusion of trauma with loss. Important for the task are Kristeva's extensions of psychoanalysis, since they show that Freud's thought, too, is inhabited by the loss of loss.

Kristeva: The Loss of an Essential Other

Kristeva's confrontation with the question of separation focuses on the early or "archaic" mother, not the oedipal father. Her thought can be read as an attempt to name, draw out, and articulate what Freud left in what he called the "dark continent": the feminine, that is, with Kristeva, the maternal feminine. Her attention to the archaic mother presents a manifold extension of the theory of primary narcissism, attentive to an early and inconclusive moment of separation that belongs to the dual relationship between mother and preverbal child. It is important to recognize that, although from an external or third-person perspective the mother's and child's bodies become radically distinct at birth, Kristeva is investigating the period of dependence on the mother when her body remains a vital necessity for the child. Yet she is concerned with the earliest moments in subjectivity when the *infans* is vulnerable, not only as a vitally dependent being, but as suddenly exposed to everything when the corporeal exchange with the mother is subject to the nonsatisfaction of infantile impulses and wishes aimed at the mother's body. Freud called this "frustration." From a Kristevan viewpoint, the satisfaction of these wishes is impossible because the mother's body cannot be an unfailing nourishing and protective landscape, for she is not mere "nature" but both the artificial extension of the child's body and a desiring and symbolic being. This is what Kristeva means by the biological-and-symbolic being of the archaic mother. Kristeva focuses less on the frustration of impulses and more on the impact of loss corresponding to the sudden exposure to everything that is brought by the dual nature of maternal being. In brief, early infantile life is itself already marked by the withdrawal of the mother's body, which is registered corporeally and psychically in drives and affects.

Kristeva's psychoanalytic thought is particularly attentive to this terrain of subjectivity where the life of the drives and affects is most emphatic because language has no hold there. In the dual relationship of mother and child drives and affects hold sway over a nonintegrated or "fragmented" body. This would not appear to be propitious terrain for discovering capacities that support the trials of separation. Classically, it is the terrain—deemed "natural"—that paternal law impels separation from. Nonetheless, when Kristeva investigates this early life of nonsymbolized drives and affects, she finds that it is a realm of preoedipal alterations in subjectivity that encompass nonlinguistic capacities to negotiate the impact of loss. Her inquiry into the fundamental nature of the loss of the other seeks out its character at the level of our scission from nature, where we are still attached to it. For this reason, she calls it the loss of the "essential" other.

The discovery of presymbolic capacities—called "semiotic" by Kristeva—undermines the view that the mother is a mere nurse and that a prohibiting paternal intervention is needed to inscribe the task of separation. The nature of the capacities that make up the early struggle with loss correspond to a condition in which there is otherness and separateness but the other remains a vital necessity. The otherness that emerges in the dual relationship of mother and child is indefinite and nondifferentiated otherness. Kristeva sets out three major alterations in subjectivity at this moment in the evolution of the individual. These are abjection, primary idealization, and primal loss, explored in *Powers of Horror*, *Tales of Love*, and *Black Sun*, respectively.[12] Together they compose a corporeal and affective—put otherwise, mimetic—responsiveness to the early exposure to loss within the dual relationship of mother and child. The most far-reaching aspect of this thought on the fundamental nature of the loss of the other appears in Kristeva's attention to the fate of this level of responsiveness in social and symbolic being. For her, corporeal and affective responsiveness must be permanent, a necessity that is explained by the complex and nuanced—for her, threefold—nature of prelinguistic responsiveness itself. The following paragraphs outline this three-dimensional nature of the loss of the essential other, which comprises Kristeva's narcissistic structure.

First, Kristeva's idea of the abjection of the mother's body presents a drive-based struggle of separation. In *Powers of Horror* this is a semiotic or nonverbal capacity on the part of the infans to attempt to throw off what is not yet parted from. The inextricably corporeal and psychical capacity of the primitive ego is that of driving out or repelling the nondifferentiated otherness that looms within the early withdrawal of the mother's body. What this capacity bears on, however, is no more than an unstable and pervious inside/outside border. As the target of drive attacks upon it, the unstable border is called the "abject." The ambiguity of the capacity of abjection is especially remarkable. For the drive attacks might represent the wish to dissolve the otherness and separateness that the emergent border brings to infantile life, yet they work as much to preserve the unstable border as to undo it. This drive-based capacity of infantile life cannot solidify the border and thereby set up the ego's boundaries, however. The abject is not and cannot become the border of the subject. Rather, abjection is a constant work of preserving the abject by displacing it. Where separation is established, the abject is discovered as the unconscious trace of this semiotic capacity in psychic life. The looming of the abject as unstable inside/outside border represents a destabilization of the ego or an ego uncertain of its boundaries: what Kristeva calls the borderline subject.

The second of Kristeva's semiotic capacities is primary idealization, a prehistorical identification with a preoedipal third. In *Tales of Love* the idea of primary idealization draws on some references in Freud to what he called the father-in-individual-prehistory. Kristeva's rediscovery of—in her terms—the imaginary father returns us to the loved father whose image disappeared from Freud's later thought on the oedipal structure. Two elements in Kristeva's rediscovery of the preoedipal third are vital for the way in which she reveals the limitations of Freud's historical reflection. First, Kristeva carefully distinguishes the imaginary father from the oedipal one. The prohibiting and judging features of the latter are missing from the former, who is nonetheless an indispensable value in the struggle of separation. This value stems from how the early exposure to loss and separateness allows for the appearance of a third that cannot settle into any position since the triangular structure that defines the oedipal conflict is lacking. The preoedipal third is neither an object of desire (the maternal apex) nor an object of fear (the paternal apex). It is a mere fleeting presence, one that is able to act as a consolation for loss by drawing to it an affective identification that Kristeva calls transference love. Her focus is not, therefore, on a field of emotional ambivalence toward the absent father. She recovers, not the loved yet dead father (in the affect of remorse), but, rather, a primary living and loving father (in the loving affect). Primary idealization is "the immediate transference toward the imaginary father, who is such a godsend that you have the impression that it is he who is transferred into you" (*TL* 41).

Second, she questions the extent to which this is a paternal instance in psychic formation. In doing so, she challenges Freud's neglect of the civilizing role of the mother in the evolution of the individual. The preoedipal third is the mother's gift, a coagulation of the mother and her desire. This is another way of drawing out the implications for subject formation of the biological-and-symbolic being of the early mother. The idea of primary idealization develops the thought that the infantile experience of the withdrawal of the mother's body brings with it the enigmatic presence of the symbolic and, especially, desiring being of the mother. The mother's desire represents her diversion from the corporeal exchange of mother and child. Nevertheless, maternal desire arises for the infans within the dual relationship and is enigmatic. The enigmatic thirdness of the mother's desire elicits a transference on to what is no more than a beyond of the dual relationship, a fleeting presence that Kristeva calls the metaphor of love. The term captures two connected notions. The first is the impossibility of the loving affect being attached to any position (loving subject, loved object, or object of desire). The second is the quality of love as being nowhere and everywhere yet imbued with otherness, such that it is the first emotional tie and thereby the support for object desire.

Once Kristeva adds this feature of ego formation to Freud's account of the evolution of the individual, she makes the gift of love the nucleus of the ego. Transference love means that the ego is founded on transcendence: an indefinite otherness that, as nonobjectal correlate of the primitive ego, supports affective relationship. It is notable that Kristeva's attention to the living and loving other recovers an aspect of religion that Freud's focus on the fate of paternal law left out. Kristeva's prehistorical affective identification has the quality of agapic love. However, this recovery is made at the cost of losing the historical import of Freud's thought on religion, and so his genealogy of the Oedipus complex, a point I will return to later. An equally important implication of Kristeva's discovery of the gift of love as a facet of primary narcissism is that it undermines the primacy of the negative connotations of narcissistic relationship. Narcissism is usually understood as an omnipotent ego unable to achieve relationship and so deprived of others. The general implication of Kristeva's metaphor of love is that separateness is not supported simply by the prohibiting and judging features of paternal law but primarily by the settling of the loving other at the core of the ego.

This means that Narcissus is a figure intrinsic to selfhood and relations with others. The capacity for transference love is, in Kristeva's view, a support for the loving idealization of otherness.[13] Nonetheless, Kristeva's attention to the gift of love is made in conditions where the suffering of ego instability encountered in psychoanalytic experience points to the weakness or absence of the dynamic of transference love.

The third of Kristeva's semiotic capacities discovered in the early relationship of mother and child turns closely on the impact of primal loss itself. In *Black Sun*, Kristeva finds that a further feature of the early life of the drives and affects can support the trials of separation, even apart from the affective relationship developed in prehistorical identifications. The third alteration in subjectivity is primal melancholy, the direct mark of the exposure to loss. The alteration of the subject, here, is a position of depression that does not manifest the loss of the object, which belongs to the realm of secondary identifications where an outside other has already turned up.[14] Instead, this position of depression calls up the "shadow of despair" cast on the emergent subject by the fundamental loss of the other before objects, properly speaking, can show up. This fundamental loss is the hinge of the whole narcissistic structure. "Narcissism would be that correlation (with the imaginary father and the 'ab-jetted' mother) enacted around the *central emptiness* of that transference" (*TL* 42). Primary narcissism is a struggle with loss unaided by the development of linguistic capacities that allow for the recovery of the mother in language. Rather, consent to loss is signaled by a mute affect: parting sadness. An affective cohesion of the self provides the unsymbolized drives and affects of the fragmented body with a preliminary integration. That is, parting sadness is a preverbal acceptance of or adjustment to the loss of the essential other. Nonetheless, once again, the prevalence of narcissistic suffering in contemporary psychoanalytic experience—the prevalence of ego instability on the couch—suggests that this affective support for mourning, too, is missing or unstable in contemporary psychic life.

As we have seen, Kristeva's attention to the early life of the drives and affects lays out a domain of preverbal representatives of the primal exposure to loss and otherness, one which amounts to a nonverbal modality of separateness and connectedness. If it is assumed that this early moment in subjectivity is a stage of psychic helplessness, where the sudden exposure to everything is a traumatic event, trauma is being confused with constitutive loss. The primal loss of the other is not an overwhelming exposure to exteriority marked by a lack of affect. It is not a trauma. Rather, it is an exposure to separateness and otherness that triggers drive-based and affect-laden capacities that underlie and are necessary to the development and sustaining of symbolic ones. We must conclude that the presence of other factors is necessary if primal loss is to turn up or be conceived of as a condition of psychic helplessness. Kristeva's attention to ego instability on the couch—the complaints about lack of love and the evasiveness of adequate selfhood—suggest that the gift of love and nonverbal capacity for loss are undermined in the secular aftermath of religion.[15]

She throws light on this condition by recognizing that the fate of the nonverbal negotiations of loss within social and symbolic life is highly significant, for the prelinguistic realm of corporeal and affective responsiveness is not simply a past in psychic life but acts as a support for adequate selfhood and connections with others. Vital for bonding, it cannot be dispensed with upon the emergence of the sign or the entrance into language. In Kristevan terms, if bonds with others are to have a "live meaning," the signs of communication, or language as communication, must be tied to the affective meaning of prehistorical identifications (*BS* 23–24). In other words, corporeal and affective relationship must be permanent if signs, bonds with others, and selfhood are not to be experienced as empty. For Kristeva, modern secular institutions and discourses neglect this need. The narcissistic suffering encountered in contemporary psychoanalysis is the symptom of this neglect. The modern Narcissus is Narcissus abandoned. This leads Kristeva to explore a maternal territory that is lost but that the subject has failed to lose. In sum, Narcissus, like Oedipus, is a figure of the loss of loss.

Before considering the wider significance of the modern Narcissus further, it must be acknowledged that there is a feature of the dual relationship of mother and child that makes the negotiation of the loss of the mother feeble at this level. In the early moment of otherness, separateness, and loss, one is suddenly exposed to everything and must do "something." This "must" is not the prohibition of paternal law but an imperative that Kristeva calls maternal authority, presiding over the preverbal struggle in which an ego attempts to come into being. In her words:

> Through frustrations and prohibitions, this authority shapes the body into a *territory*, having areas, orifices, points and lines, surfaces and hollows, where the archaic power of mastery and neglect, of the differentiation of proper-clean and improper-dirty, possible and impossible, is impressed and exerted. . . . Maternal authority is the trustee of that mapping of the self's clean and proper body; it is distinguished from paternal laws. (*PH* 72)

Maternal authority presides over a not yet stabilized inside/outside border, where there is scission from nature, still attached to it. It therefore presides over the shaping of a body. The crucial point for our purposes is the ambivalence of maternal power: how it is a problem precisely because it is wholly played out in the shaping of the body. On the one hand, this is a form of authority, which Kristeva calls primal repression in order to distinguish it from the—now "secondary"—repression of paternal laws. Maternal territory is a form of authority because it has a notion of correctness: a demand for separation. Kristeva's presymbolic alterations in subjectivity are vital for the possibility of separation understood as the subjective autonomy developed in and sustained by language or the world of signs. The appearance of social and symbolic capacities therefore requires the presymbolic life of the drives and affects—their modality of separateness and connectedness. On the other hand, and herein lies the problem, the success and failure of maternal authority are not clearly distinguishable. The mother's power is intrinsically unstable, not because it lacks subject-object form—that is, discursive or objectal form—but because it is all played out on one spot. In other words, in this moment of separation, where there is only indefinite and nondifferentiated otherness, the various moments of separateness and connectedness remain indistinct. The imperative to separate (maternal authority) what it calls for (the repulsion of the maternal vessel) and where it is carried out (the shaping of a body) are indistinct moments because the site of maternal authority is the still nondifferentiated terrain of the mother's body and the child's body. This is what makes maternal power an overinclusive and overbearing authority. The successful form of maternal authority carries with it the mother's overauthority, for the struggle of separation from her occurs within the vital dependence on her. This early moment of separateness therefore carries with it the need of another form of separateness at a different level, classically our "oedipal destiny" or the appearance of paternal laws.

There is a further need, however, as Kristeva has argued, for this second form of separateness to preserve the corporeal and affective relationship that begins with preoedipal relationship. Otherwise the social and symbolic capacities enabled by the second moment of separateness will be disabled since they will not encompass bonds with others. In other words, all three of Kristeva's presymbolic alterations of subjectivity—loving idealization (the affective constitution that settles transcendence at the nucleus of the ego), sadness (the fragile cohesion of the ego gained in and through the nonverbal acceptance of loss), and the drive-based repulsion of the other (a primitive moment of separateness)—must have a living history because they are necessary for bonding. If they are cut off from symbolic life, symbolic and social capacities are uprooted from a source that makes them valuable and meaningful. In other words, the prevalence of symbolic capacities in which semiotic capacities have no vocation is a merely linguistic universe, one without a world of nature and so without history. Taking the reverse perspective, without the second moment of separateness, the presymbolic alterations in subjectivity remain just that: presymbolic, inchoate, or "split off." The need of permanent corporeal and affective relationship is the need of repeated and successive transformations of nonverbal love and loss, carrying them into the life of signs. These transformations shift corporeal and affective relationship off the spot of maternal territory, so to speak. Kristeva's diagnosis is that there is a dearth of social and symbolic mediations of corporeal and affective responsiveness in the secular aftermath of religion. This notion is fundamental to her project in *Powers of Horror*, *Tales of Love*, and *Black Sun*. Her own thought develops the ways in which certain practices in modern life have stepped into the gap left by modern secular discourses and institutions. These are Kristeva's imaginary discourses: practices, usually artistic at this point in her writing, that achieve the rebinding of symbolic constructions with semiotic traces. For example, she reads certain works of art and literature as tales of love or works of mourning, suggesting that they give a life to the neglected realm of affective and corporeal responsiveness. That is to say, in modernity the values of love and loss for selfhood and connections with others are recovered in the realm of semblance or artifice. Kristeva's reflection on the need of imaginary constructs also leads her to develop ethical ideas that recover the import of embodied experience for separateness and connectedness. These ideas counter the abstractedness of moral thought, notably Platonic and Kantian, which comes from founding the ethical in a rationality that turns against the passions or impulses. Kristeva's imaginary discourses are not the recovery of any raw site of drives and affects, however. They are symbolic mediations of affective and corporeal responsiveness. This dimension of Kristeva's thought cannot be followed through further here, since we are on the track of how the confusion of trauma with constitutive loss appears in late modern societies.[16]

I suggest, first, that the prohibitive function that plays the role of the second moment of separateness in Western cultures—oedipal destiny—does not resolve but, rather, exacerbates the problem of the neglected Narcissus. For the oedipal structure of subjectivity is one in which affective bonding is undermined. This is the problem that turned up in Freud's thought in *Civilization and its Discontents.* However, when Kristeva follows through on Freud's neglect of the maternal role in the trials of separation, she reveals the limitations of his own historical reflection. The prohibitive function does not only represent the repression of desire. It weakens the possibility of negotiating the loss of maternal territory in ways that would preserve, through mediation, the realm of corporeal and affective responsiveness. The oedipal structure of subjectivity is equally the neglect of the corporeal and affective dimensions of subjectivity. Nonetheless, the investigations of the oedipal and narcissistic structures of subjectivity offered by Freud and Kristeva, respectively, have revealed that the constitutive source of separateness and connections with others is loss. When our thinking confuses constitutive loss with trauma it retains all the features of a subject that inhabits the loss of loss, one suspended between the overstrictness of paternal law or superegoic morality, on the one hand, and the overbearance of maternal authority, discovered in the abandoned modern Narcissus, on the other. This is a subject caught between a rock and a "no-place."

Conclusion: The Historical Confusion of Trauma and Loss

What remains to be done is to confront the problem that a deus ex machina has been operating in the discussion hitherto. For no account has been given of how, in Freud, a past social reality, the authority of religion, could become a psychical reality: the unconscious guilt structure of subjectivity. It might be tempting to think that the way to go is to unify the two stories of the genesis of subjectivity. One overarching story that successfully tied up the various aspects of the loss of loss turned up in these investigations of Oedipus and Narcissus might throw some light on the two remaining questions: What leads a social reality to become a psychical reality? How can trauma and constitutive loss come to be confused? Such a story would need to set the major discoveries about Oedipus and Narcissus in the right kind of relationship to each other. When Freud connects psychoanalysis and religion, he underlines that the repression of desire persists through the oedipal structure because the latter is the trace of an absent religion that had forgotten that the loss of the other is the constitutive source of its authority. This needs to be put into relation to Kristeva's discovery of the repression of the maternal feminine and corresponding neglect of the corporeal and affective dimension of bonds with others. There is some appeal in the idea that the recovery of the feminine is itself a process that goes over and weakens the repression of desire. Indeed, the appearance of the notion of *jouissance* in French thought—the idea of a pleasure that exceeds the strictures of symbolic paternal law—might suggest that such a process is under way. There is also some appeal in the thought that uncovering the role of corporeal and affective relationship in bonds with others contributes in and of itself to overcoming the severity of paternal laws.

Despite this, the two versions of subject-formation that Oedipus and Narcissus represent cannot be unified into one overarching story. They cannot take account of or get on level terms with each other. On the one hand, paternal law does not account for maternal authority. The archaic mother is and must be biological-and-symbolic if the early mother is not to lose the features through which she is known as, not a naturalized body, but a civilizing role. Even though this role rests on the necessarily symbolic being of the archaic mother, she is not the so-called phallic mother—not, that is, an early moment of symbolic, paternal law. On the other hand, maternal authority does not account for paternal law. In order for it to do so, this first form of authority would have to underlie the second one as its condition of possibility, and the second form would have to be the later moment of separateness that not only shores up the first moment but allows corporeal and affective responsiveness to gain a life in the social and symbolic realm. This is not so, as the Freudian investigation of paternal law has shown. The second moment of separateness inscribes the subject in a guilt structure and undermines social being. The most that can be said is that there are two histories of subjectivity. That is, as indicated at the beginning of this chapter, Freud and Kristeva offer two distinct genealogies of subjectivity that cannot be unified. This also means that the two questions I posed at the beginning of this section about the development from social to psychical reality and the confusion of loss with trauma remain unanswered.

Perhaps neither one of the two distinct genealogies of subjectivity has succeeded in fully addressing our inability to think through the fundamental nature of the loss of the other, so that neither can really illuminate the confusion of trauma with loss. Freud did not interrogate with sufficient self-reflection the conditions under which a past social reality, traditional authority, could become an exclusively psychical reality: the unconscious sense of guilt. Moreover, Kristeva's investigation into the maternal feminine may reveal the limitations of Freud's historical reflection but it also appears at the cost of losing his genealogy of the Oedipus complex. This is why the questioning of trauma cannot be abandoned. If the conception of trauma keeps turning up in the wrong place, as an originary moment of subjectivity, or the return of the other, or a condition of futurity, it needs to be asked what the connection might be between the problem of the loss of loss, on the one hand, and the thinking of trauma, on the other. Put otherwise, if trauma is not original, what is it "for us," such that we make this mistake? The confusion of loss and trauma is not a mere cognitive error. It has substance. The two outstanding problems—both the possibility of confusing trauma with loss and the need of removing the deus ex machina in the notion that religion begat the psyche—can be approached by considering whether they might be one problem.

To say that the confusion is not a merely cognitive error is to say that its substance lies in the process whereby a past social reality has not passed by but, rather, has become a psychical reality. Psychoanalytic thought on trauma throws light on the historical meaning of "internalization." We saw in Freud's account of the Exodus how a break from the past can be a traumatogenic event: a situation of danger that is not itself experienced, with the trauma awakening when the danger is already past. With Freud, traditional authority—the authority of religion—is the working out of the traumatogenic appearance of a call to morality and justice whose essence is prohibition. The failing of religion is the loss of its constitutive source in the loss of the other, and this loss of loss is continued anew in modern subjectivity, where the power of the absent father is a psychical and not social reality. Light can be thrown on the confusion of trauma with loss by raising the question, What is it for such a failure to be continued anew? I am asking about the general nature of and relationship to the rise of modernity. It can be assumed that the passing of traditional forms of life brings with it the release from traditional authority. Yet, on the Freudian view, this is not what has happened: the psychic life of modern subjects exhibits the persistence of a past authority, still suffered in its absence. In other words, the release from the past is not undergone, and the binding of past, present, and future is undone. Discontinuity and contingency are oft remarked upon features of the temporal structure of modernity. But what is in view here is the nature of the entrance into modernity. Given Freud's discovery that the release from traditional authority, given by the passing of traditional forms of life, does not bind that authority into the past, modernity itself appears to be marked by an event that has not passed by. That is, the entrance into modernity is a traumatogenic event: not itself experienced, and so not historical. The idea that a past social authority becomes a psychical reality—that the failings of traditional authority persist in modernity in an internalized form—is only fully comprehensible on these grounds. The past juts into the present in a split-off form, that of the guilt structure of subjectivity. In sum, insofar as Freud's "discovery" of the unconscious turns constantly on the oedipal structure, it is an exploration of the repercussions of the traumatogenic rise of modernity. If his genealogy of the Oedipus complex is lost, and the oedipal structure is considered to be universal and transhistorical, then it will appear that he puts trauma at the origin of subjectivity as such. This appearance is overcome once it is recognized that Freud's Oedipus is not only modern but a figure of the entrance into modernity as an event that is not itself experienced.

In conclusion, the discovery of trauma in subjectivity is the discovery of an historical not an original structure. I have shown, then, that psychoanalysis opens up a genealogy of trauma. The thought on subjectivity that puts trauma at its origin is a failure in self-reflection, for what needs to be recognized is that the idea of original trauma is a projection. It represents the failure of the thinking modern subject to recognize itself in the object of its thinking. The subject who inhabits the loss of loss is the modern subject, suffering the ruin of experience this condition imposes, preoccupied by its prehistory, and unable to go over the rise of modernity or bind the past into the past. The appearance of the thought of original trauma can be viewed as an attempt to repair this suffering by returning an—illusory—experience to the modern subject. It is trauma that is to do the work of binding past, present, and future. That is, trauma itself is made futural. This attempt to turn the source of suffering into the relief from it may have some logical appeal for those who grasp the temporal peculiarities of trauma. Nonetheless, succumbing to the appeal of logic prevents one, as Freud warns in another context, from asking whether a similarity in structure allows us to conclude that trauma and immemorial loss are identical in essence, or, better, in substance. In other words, succumbing to the appeal of logic does not illuminate how the immemorial past, and so the fundamental nature of the loss of the other, is negotiated rationally and historically.

The predicament of subjectivity that Freud and Kristeva investigate—the loss of loss—suggests the following. First, our ability to think through the fundamental nature of the loss of the other depends on the loss of the mother. That is, it requires the loss of the foundation of a body—a loss without which the body cannot live—rather than the disidentification with the mother that the paternal function imposes.[17] Kristeva's analysis actually shows that the modern fate of sexual difference is structured by the loss of loss. Second, thinking through the fundamental nature of loss requires the Freudian discovery that when religion or conscience are made foundational for culture, the magical, taboo structure of authority is maintained. Recognition that the foundation is an illusion is necessary if release from it is to become possible. Finally, the thought of original trauma, allowing trauma to be the binding of past, present, and future, is a tempting one, not least because it can act as a consolation for our losses. Nevertheless, it is a misstep that repeats the inability to think through the fundamental nature of the loss of the other. It is a further, highly compacted, symptom of the perennial historical crisis of modernity.

Notes

1. On nonpsychoanalytic ground, Levinasian ethics introduces a conception of trauma as an immemorial event that uproots subjectivity from its narcissistic egoism and returns it to the Other. One might say, the other turns up again in trauma. Trauma then becomes central to the Levinasian conception of ethics as responsibility. See Emmanuel Levinas, "God and Philosophy," in *Basic Philosophical Writings*, ed. Adriaan T. Peperzak, Simon Critchley, and Robert Bernasconi, (Bloomington: Indiana University Press, 1996). When the attempt is made to connect this thought with psychoanalytic thought on trauma—as it is, for example, by Simon Critchley in his essay "The Original Traumatism: Levinas and Psychoanalysis"—it is already assumed that when psychoanalysis puts an immemorial exposure to otherness in the early moments of subject-formation this is a traumatic event. With Critchley, the subject is constituted in a transferential relation to an original trauma, and this is the condition of possibility of the disposition toward alterity, and so of any ethics of phenomenology or ethics of psychoanalysis. See Simon Critchley, *Ethics-Politics-Subjectivity: Essays on Derrida, Levinas, and Contemporary French Thought* (London: Verso, 1999), 184-85.

2. J. Laplanche and J. B. Pontalis, *The Language of Psychoanalysis* (London: Karnac Books, 1988), 286.

3. Sigmund Freud, *Totem and Taboo* (1912–13), in *The Standard Edition of the Complete Psychological Works of Sigmund Freud*, ed. and trans. James Strachey (London: Hogarth Press, 1953), 13:141–42. Hereafter cited as *TT*.

4. Julia Kristeva stresses that this is an identification with the function of authority in *The Sense and Non Sense of Revolt: The Powers and Limitations of Psychoanalysis I* (1996), trans. Jeanine Herman (New York: Columbia University Press, 2000), 44–45.

5. Sigmund Freud, *Moses and Monotheism* (1939), in *The Standard Edition of the Complete Psychological Works of Sigmund Freud*, ed. and trans. James Strachey (London: Hogarth Press, 1953), 23:60. Hereafter cited as *MM*.

6. "Ordinarily God and people are indissolubly linked, they are one from the very beginning of things. No doubt we sometimes hear of a people taking on a different god, but never of a god seeking a different people" (*MM* 45).

7. Freud proposes that, as the exiled tribes combine with others on their journey, the God they worship is not that of Moses but a volcanic God, Yahweh, whose characteristics—"a coarse, narrow-minded, local God, violent and bloodthirsty"—are at odds with those of the One God who demands faith and a life in truth and justice (*MM* 50).

8. Julia Kristeva, *Strangers to Ourselves* (1988), trans. Leon S. Roudiez (London: Harvester Wheatsheaf, 1991), 181.

9. See Gregg Horowitz, *Sustaining Loss: Art and Mournful Life* (Stanford, CA: Stanford University Press, 2001), chap. 4.

10. Freud's detailing of the reassertion of the Mosaic doctrine underlines that the return of the repressed does not mean that the murder of the founder itself is remembered in Hebraic monotheism. He suggests that it is represented—not recalled—in the doctrine of original sin and the idea of the death of Christ, in which the Son who takes the guilt of the world on his shoulders. This is, in Freud's view, a "phantasy of expiation" in respect of "a shadowy 'original sin'" (*MM* 135). The murderous deed for which expiation is required remains unnameable. That is, the salvation religion is a further elaboration of guilt culture.

11. Sigmund Freud, *Civilization and its Discontents* (1930), trans. James Strachey (London: Norton, 1961), 89.

12. Julia Kristeva, *Powers of Horror: An Essay on Abjection* (1980), trans. Leon S. Roudiez (New York: Columbia University Press, 1982); *Tales of Love* (1983), trans. Leon S. Roudiez (New York: Columbia University Press, 1987); and *Black Sun: Depression and Melancholia* (1987), trans. Leon S. Roudiez (New York: Columbia University Press, 1989). Hereafter cited, respectively, as *PH*, *TL*, and *BS*.

13. This may be the idealization of actual others or the idealization of some symbolic otherness such as a cultural value or a meaning. Indeed, when she speaks of idealizing constructions in *Tales of Love* we get the sense, developed in later works (*Strangers to Ourselves* and *Nations Without Nationalism*), that what is crucial is the combination of the two. The idealization of others supported by symbolic values prevents idealization of an other

from threatening an abdication of the ego and its judgment. Conversely, the idealization of cultural values and meanings mediated by actual others prevents symbolic life from becoming a mere linguistic universe disconnected from others. Idealizing constructions are therefore crucial for the possibility of the social bond in Kristeva's thought.

14. For object loss in the realm of secondary identifications, see Sigmund Freud, *Mourning and Melancholia* (1917), in *The Standard Edition of the Complete Psychological Works of Sigmund Freud*, ed. and trans. James Strachey (London: Hogarth Press, 1953), vol. 14.

15. Kristeva concludes *Tales of Love* with the thought that "the lack of a secular variant of the loving father makes contemporary discourse incapable of assuming primary identification—the substratum for our idealizing constructions." "The fact that today we have no love discourse reveals our inability to respond to narcissism" (*TL* 374, 381).

16. For a reading of Kristeva's thought on the symbolic mediations of the semiotic, see Sara Beardsworth, *Julia Kristeva: Psychoanalysis and Modernity* (Albany: State University of New York Press, 2004), chaps. 6–8.

17. If the primacy of the oedipal structure is maintained, it is assumed that social and symbolic being is won through disidentification with the mother. This is to make misogyny structural for culture as such. Kristeva's undermining of this implication on psychoanalytic ground therefore joins the manifold feminist investigations of a subjectivity historically (not essentially) structured by disidentification with the mother and the repression of affective relationship: from de Beauvoir's interrogation of the woman question, to Gilligan's and Chodorow's respective inquiries into the development of gendered personality, and the thought of care ethics, to recent French feminism's investigation of sexual difference.

4

Trauma and Speech as Bodily Adaptation in Merleau-Ponty

Kristen Brown Golden

Introduction

At a concert recently, my friend Rita stood for the entire second half due to a cramp that stretched from her thigh to her lower torso. During the intermission Donna, a physical therapist, cautioned Rita to stop her regular exercise routine for a couple of weeks. "Otherwise," Donna explained, "the associated neurons and muscles will learn to remember for good the painful and unhealthy way they are coordinating now." Many of us living in industrialized societies of the twenty-first century find ourselves, like Donna, speaking in surprising ways about the relationship between human learning and memory, and human physiology. Just what do we mean when we say, "muscles learn and remember?" In the seventeenth century, René Descartes articulated a theory of the relation between so-called mind and body. The mind, or an idea, he said, is separate and distinct from the body or brain with which it associates. One deep-seated conviction in the West influenced by the Cartesian legacy is that thought (traditionally including language, memory, and learning) happens above the neck, and whether or not "in" the brain (a matter of debate)—at least not below it.

Donna's statement to Rita, interestingly, harbors an implicit challenge to the Cartesian legacy: neurons and muscles below the cerebral cortex can learn and remember. Indeed, recent spinal cord research supports this view.[1] The question of mind-body relation, then, is a question about boundaries—not merely about respective boundaries of mind and body but also about those of any item or being. Where does one "begin" and the other "end?" Do we know? And if we do not know, in what ways does a loss of belief in discrete entities, or identities matter?

Even as physical therapists today invoke the language of muscle memory, for most of them (and us), our modern speech and perceptual structures keep our awareness bound to Cartesian concepts. Many of us may accept an analogy between muscular learning and memory and so-called ordinary human learning and memory, but will assume that these are categorically different—that the difference between human thought and muscle memory is more than a degree of complexity. Indeed, a prevailing unexamined assumption is that at their root, human reflection and physiology are fundamentally different. The influence of this view reflects how entrenched are the following three contemporary views that descend from modernity and mutually imply one another. These are the widespread conviction in Cartesian mind-body separateness (or dualism), belief in Enlightenment logic of Galileo-Newtonian mechanistic cause and effect, and commitment to the techno-scientific assumption of separateness of subjects and objects, selves and others, beings and things.

Through a reading of the writings of phenomenologist Maurice Merleau-Ponty (1908–61), this chapter exposes as misguided such a Cartesian legacy through an examination of several contemporary practices and experiences. Merleau-Ponty is known among phenomenologists for emphasizing and privileging the role of the body in human experience.[2] For Merleau-Ponty, animal corporeity is constitutively communication,[3] which ranges from the primordial (an eye moving and perceiving movement as a kind of interrogation and response) to the complex (humans speaking and engaging in higher order cognition); communication is rooted in corporeity that directs itself outward to delineate itself and make of itself a "presentation."[4] Speech, I shall argue, when understood to be an expression of its origin—corporeal desire—shows itself to be less a mode of representation than a mode of bodily adaptation.

That communication is, as I show, rooted in corporeity proves important for several reasons. It shows the body as a whole, displaying activity traditionally believed confined to the brain and to cognition. It implies a substitute for the traditional hierarchical chain of being—and its correspondent differences in ranks of beings: a lateral plane of beings sharing equal rank. Finally, by showing that corporeity exists as communication, but that communication is a bodily constituent, my view is for an additional reason crucial and important. It shows as connected that which has often been seen as disconnected: idea and matter, mind and body, psychology and biology.

While many have critiqued mechanistic models of self and being entailed by a Cartesian mind-body dualism, some using the philosophy of Merleau-Ponty, my examination is unique in how it strengthens such a critique: through recent empirical research on physical and psychological trauma. More specifically, it analyzes lived experiences of traumatogenic narrative speech—in the aftermath of war or domestic violence—and research on adaptive anatomy—in the aftermath of spinal cord injury.

Rarely have Merleau-Ponty's writings been placed in dialogue with contemporary literature of trauma studies.[5] Virtually none of the philosophical commentary on Merleau-Ponty acknowledges a concept of trauma implicit in Merleau-Ponty's work.[6] Nor does such Merleau-Ponty criticism recognize how an approach to trauma operative in current trauma research—that is, that of not only psychological but also physiological trauma—supplies new evidence for Merleau-Ponty's view of humans and nature, bodies and psyches as interpenetrating.

Language as/of Corporeality

All perceptual structure (i.e, that of space, time, sight, sound, smell, taste, touch, and abstract reflection) can be said to begin for Merleau-Ponty as and from corporeity. Moreover, corporeity exists as a structure that mediates all structures. For Merleau-Ponty, organs such as skin, eyes, and veins live a primordial language (*N* 222). In his lecture courses on nature Merleau-Ponty uses the expression "natural" with respect to language to connote meaningful gestures (raising an eyebrow, nodding one's head). Cecilia Sjöholm notes that the term "natural" language denotes an "even more primordial" sense of language.[7] "A sense organ like an eye or a hand is already a language because it fills the function of interrogation and/or response. What Merleau-Ponty calls interrogation, here, is movement and the response is perception, a response to movement" (*EM* 175). A sense organ's perceptive response could be said to signify to surrounding organs the movement it has just perceived. "The body is, in other words, already in a situation where it communicates with itself: touching itself, responding to itself, as well as to other bodies" (*EM* 175). All but the most simple animals, then, capable of responding to their environments—and not primarily reacting as in the case of the simplest animals[8]—can be said to be a corporeal communication. Such corporeity, then, appears to exist as communication.

Significant here is not only that corporeity exists as communication but that it does not exist of communication. Put differently, the communication corporeity exists as is not the basis of corporeity, rather, corporeity is the basis of communication. Without a body and nervous system, then, signification (or communication) appears unable to arise. Thus, communication shows itself rooted in bodily functioning (not vice versa).[9]

Whether primordial or complex, any communication "can never be wholly detached from the function of the body," and all communication has roots in the body (*EM* 175).

Corporeality as Signifying Self and Other

Animal bodies are also the basis of communication for Merleau-Ponty in another sense: a body's surface functions as a communicative sign. It is a signifier that signifies "oneself."[10] For Merleau-Ponty, corporeity shows itself to be impelled toward self-signification.[11] He shows animal bodies negotiating their surfaces to signify themselves, in his analysis of the work of biologists J. von Uexküll, E. S. Russell, and Adolf Portmann (*N* 167–90).

Vital behavior "is an oriented action, it points itself to something, and this pointing, this indication, is already signification; vital behavior, in virtue of its being-oriented-to, thus *signifies*" (*IJ* 198). In *The Structure of Behavior*, Merleau-Ponty explains that a vital organism's attribute of signifying is "an attribute of the *perceived* [and perceiving] organism" (quoted in *IJ* 198). The perceived and perceiving organism, according to J. von Uexküll, includes all but the most simple of animals (*N* 168–173).

J. von Uexküll's work is significant for Merleau-Ponty because it shows that many animals with nervous systems respond to and orient themselves within their environments. Unlike the "lower" animals—such as worms and sea urchins—they appear opened-up to their environment (*Umwelt*) rather than closed to it. The lower animals, by contrast, seem not to relate to their environments. Instead their behavior seems governed according to an anatomical construction (*Bauplan*) that functions much like a machine (*N* 168). So-called higher animals, on the other hand, are characterized precisely by their relating/orienting themselves amid the environment. Moreover, higher animals respond to the Umwelt as if it were a world of signs (*Merkwelt*). By focusing on the development of the retina in animals, Uexküll shows that "the agency of the exterior world, the objective universe, from now on plays the role of a sign rather than that of a cause" (*N* 171). The nervous system operates according to Uexküll as a means for interrogating, organizing, and responding to forces of the external world (*N* 171). What is important about Uexküll's work for Merleau-Ponty is that all but the most simple animals are seen as moving not mechanistically but instead as creatively and in response to signs. This suggests the crucial distinction in Uexküll's concept of environment. An environment "is not reduced to a sum of exterior events" but is cocreated by the conditions of and the animal's regulations to the milieu (*N* 177). In this perspective, behavior of all but the simplest animals is a receiving, from the world,[12] of signs from among which it chooses and which it shapes to its own tendencies, its environment (*N* 177).

Merleau-Ponty shows E. S. Russell's work complementing the significance of Uexküll's. Russell discusses the movements of cells in the case of a surface wound. Cells move from deep in the body toward the wound, from the interior toward the exterior—toward the body's beyond. Animal life differentiates its interior from its exterior and beyond with a "directiveness." The implication is a "weak teleology," weak in the sense that it involves a "nonfinality," that is, no determinate end-objective (*IJ* 200–201). Its open-ended finality is a purposive mobilizing from animal interior to animal exterior—even beyond the animal's body. Russell's work is important for Merleau-Ponty because it shows the formation of a condition necessary for the possibility of signification: the animal's negotiated body as differentiated from a milieu "beyond" it. This indicates a "sort of pre-signification."[13] Reference to an exterior presupposes a sense of "cohesive unity of the animal body" (*IJ* 201). The animal body has to have marked differentiations that can in turn "signal" or "point to" themselves and others as themselves and others—whether as itself as an animal whole, or as other external to itself.

If, for Merleau-Ponty, Russell's work helps him show that animal life as differentiation provides a precondition for the emergence of animal life as signification, Portmann's work displays such signification's emergence. Portmann stresses the importance of the perceivable surface of an animal body. Whereas Russell shows that animal life marks differentiations and shows a directiveness from inside to outside, pointing beyond the body, Portmann reveals the significance of the animal body as an organ for expression and for being recognized (*N* 187).[14] Portmann attributes to this ability what he calls a "value of form" (*N* 188). Merleau-Ponty interprets this to mean that animals' ability to show themselves as a designated whole to others, especially others of their species, may have value in itself; that is, it allows perceivers viewing the whole of the animal to behold the "mystery of life" (*N* 188). "The form of the animal," says Merleau-Ponty, "is not the manifestation of a finality, but rather of an existential value of manifestation, of presentation" (*N* 188).

The self-showing of animal corporeity has an "expressive value" (*N* 188) as "an organ for being seen."[15] In its appearing, the animal enacts a signifying—it signifies itself, differentiating itself from its surroundings and other creatures. Merleau-Ponty's analyses of Uexküll, Russell, and Portmann suggest that animal behavior is thus constitutively a process of communication as self-signification. Animal behavior is oriented to creating a sign that signifies itself. The sign for itself that animal behavior shapes is the animal's own corporeal surface. Uexküll's research provides a concept of animal environment that assumes that an animal perceives a world of signs and regulates them to coconstruct its environment with conditions thereof. Russell's research complements Uexküll's by showing animal behaviors that precondition self-signification: the processes of differentiating inner and outer. Portmann's view of the animal body surface functioning as a sign—a sign of itself to be seen by others—shows the emergence of self-signification. Seen in a context combining the research of Uexküll, Russell, and Portmann, the body's surface can no longer be viewed as merely a static or given sign, but rather one that is hard won and actively differentiated by the animal itself from its surroundings.

The animal's negotiated surface is that aspect of itself that is perceivable by the perceiving others, which become, at least in part, differentiated from it through the layered process of self-signification I have described. Animal behavior in this context can be said to manifest communication not only because it perceives signs that it interrogates and negotiates but also because it signifies meaningfully: it directs itself outward and makes of itself a signifier. The signifier is meaningful because it communicates a provisional externality of the animal. The signifier implies, "This is me. This is my territory." It furnishes a necessary condition for subsequently pointing to and identifying conditioned others. In this way, bodies can be said to exist both as communication and as self-signification.

Merleau-Ponty indicates self-signification as basic to a corporeal structure that constitutes the lived adaptive experience of all but the simplest animals.[16] What is significant about this notion that all but the simplest animals live oriented toward a self and other designation is that it shows corporeity as the root of communication, and demonstrates that existing is itself communication.

Physiological Adaptation

Thus far I have emphasized Merleau-Ponty's analyses of animals. These focus on nonhuman animals but inferentially extend to humans. In *Phenomenology of Perception*, Merleau-Ponty's analyses emphasize, by contrast, humans. There Merleau-Ponty appeals to cases of human brain injury, limb amputation, and paralysis, among others, and implies the exceptional status of trauma for his project. He does not, however, explicitly explore, specify, or reveal his concept of trauma. Here I will begin such an analysis. I will explore three kinds of bodily adaptation: physiological, technological, and verbal—as expressions of preconscious bodily desire and especially of that which is arguably a human being's most intense kind of preconscious desire—that generated by traumatic experience. By focusing on trauma and its corollary—bodily desire at its most extreme—I show the remarkable extent to which Merleau-Ponty is able to make corporeality visible as directed toward a self-signification that both expresses and sinks its roots in corporeity.

Persons injured during combat typically experience and respond to trauma across several interpenetrating levels of embodied adaptation.[17]

I will provisionally divide these into three and call them physiological, technological, and verbal adaptation. Merleau-Ponty's work generally offers little developed analysis of what I call "physiological adaptation."[18] By physiological adaptation I mean the spontaneous mobilizing of an organism's body cells, tissue, and organs, in response to preconscious felt needs. If physiological adaptation is rarely scrutinized in *Phenomenology of Perception*, it is nonetheless alluded to repeatedly via Merleau-Ponty's references to amputated limbs, injured brains, and paralyzed bodies. The analysis of physiological adaptation that I offer here is my own. I limit the discussion to one example of traumatized body to which Merleau-Ponty refers—paralyzed body. In particular, I will focus on recent research on the spinal cord–injured body.

In cases of paralysis from spinal cord injury, the physiological body marshals a complex set of internal adjustments to strike an equilibrium with its new circumstances. Among persons with cervical spinal cord injuries, that is, above the spinal segment range of thoracic one to lumbar two or three (T1 to L2 or L3), a process of equilibrating unfolds over a period of weeks, months, and years. The spinal cord injury alters the functioning of many corporeal processes and one such process is the autonomic nervous system—the part of the vertebrate nervous system that governs automatic internal processes (*SI* 32–33). Spontaneous processes such as cardiovascular function, internal temperature regulation, secretion (sweat, saliva, urine), and intestinal contraction (enabling excretion) begin operating under the extremely altered circumstances caused by the injury (*PAF* 451–55). Within the first several weeks after the injury, persons with injuries above the T1–L2/L3 spinal range (an area that governs the sympathetic nerves of the autonomic nervous system) are prone to experience bradycardia (slowing of the heart), bradyarrhythmia (alteration of heartbeat) and hypothermia (*SI* 32–33).

Significant in my view is the adaptive process of the sympathetic and parasympathetic nervous systems within weeks of the injury. After about three weeks, spinal cord–injured persons stop experiencing an extreme tendency toward hypothermia. This is because of a dual adaptive process. The sympathetic spinal nerves begin to function again—although at a reduced level—despite broken communication between them and the parasympathetic nerves. Also, at about this time, the parasympathetic nerves begin to reduce their output, causing a disproportionate dilating effect to compensate for the effect of constriction given the compromised sympathetic nerves. A gradual process of physiological shifting sets in. After the injury, one's spinal nerve physiology shows itself vigorously exploring and attempting to survive the new circumstances. Through a process of interrogation and response, the autonomic nerve processes of the spinal cord–injured body gradually reorient themselves and adjust. Such adaptation could be said to exhibit what Sjöholm calls "primordial 'natural' language" (*EM* 175).

By moving (interrogating) and perceiving (responding), the compromised autonomic nervous system creates a new strategy for temperature regulation, given the altered conditions. If it succeeds, it will have equilibrated itself relative to its new environment. While persons with spinal cord injuries remain more susceptible to both hypothermia and hyperthermia (due to loss of sweat glands)[19] even after the autonomic nervous system adaptations I have described, they are much more comfortable and thermodynamically equipped to survive than they were during the first three weeks following the injury. Significant for our argument is that the autonomic nervous system shows the capacity of mammalian physiology to negotiate its place and reconfigure itself according to the new circumstantial demands.

If the traumatized autonomic nervous system makes its most steep show of adaptation during the first three weeks after the injury, its adapting and that of other physiological processes continue in noticeable ways for roughly the first two years after the injury. Generally during this time persons perceive a gradual increase of overall bodily ease of functioning and comfort. Several years after his accident Christopher Reeve's body, though still more susceptible to heat and cold than a non-spinal cord–injured body, and barring other complications, had become more efficient and felt for Reeve more comfortable than during the early months after the injury.

Skin adaptation is another example of physiological shifting and a consequence of trauma to the spinal cord (*SI* 32–33). Here, as in the function of temperature control, the threat is strongest during the first months after the accident. While skin resilience and adaptation is unique for each body, it is common knowledge among the spinal cord–injured that, after several years, the live tissue of some paralyzed bodies adapts to sitting or lying for many more hours in one position in a wheelchair or bed than it initially could do. In some cases, bodies of persons with spinal cord injury have transformed part of their live tissue into thick dead callus, in combination with certain other internal tissue adjustments, to defend against the internal spread of dead tissue.

While the self-demarcating, reorienting aptitudes of more simple anatomical mechanisms such as autonomic nerves and skin tissue may be less expansive than those of more complex mechanisms (cerebral neurons), they are also more expansive than has been traditionally believed. As James Grau and Robin Joynes show in recent studies about Pavlovian and instrumental conditioning within the spinal cord, the spinal cord neurons can learn and remember, albeit in simple and more limited ways than cerebral neurons.[20] Grau and Joynes's conclusions are based on experiments with rats whose spinal cords have been transected (*PC* 13–54).

Even if the spinal cord behaves and learns more simply and with more constraints than the brain does, Grau and Joynes note that its aptitudes have nevertheless been underestimated. According to traditional views, learning requires consciousness and/or the more complex networks and neurons of the brain. In other words, learning has been believed to happen above the spinal cord, not in it or in other noncerebral anatomy. Studies by Michael Patterson (about spinal fixation in animals),[21] Jonathan Wolpaw (about spinal stretch reflex),[22] and Anton Wernig, Andras Nanassy, and Sabina Müller (about treadmill therapy for paraplegics and quadriplegics)[23] further support the notion that spinal cord neurons learn and remember. Such developments in spinal cord research are significant for Merleau-Ponty studies. They corroborate Merleau-Ponty's view of animal behaviors as interpenetrating coresponses to felt bodily needs, not self-contained faculties with defined functions.

Examples of spinal neuron, autonomic nerve, and tissue adaptation in spinal cord injured–bodies are significant because they show physiological processes behaving as communication—interrogating and responding to felt needs. They are reminiscent of Sjöholm's view (*EM* 175) that a form of primordial language is implied by Merleau-Ponty when the latter analyzes the movement (interrogation) and perception (response) of cellular physiology (*N* 219). The extreme cases of spinal cord trauma also evince physiological processes as an organism's orientation toward self-signification. They support the idea that animal behavior for Merleau-Ponty exists as communication, with a weak teleology aimed toward self-designation or the making of oneself a "spectacle" to be seen (*IJ* 200). We see this desire to negotiate with one's surroundings a territory of "self," in the gradual processes undergone by the spinal cord–injured. Physiology begins to respond to altered conditions for temperature control and maintenance of live tissue. Physiological adaptation in the case of extreme trauma reorients and recreates physiological relations in the face of new circumstances. The gradually emerging, newly equilibrated physiology of the spinal cord–injured person underlines the idea of physiological motility as a self-showing, differentiating self from other.[24]

The existence of this designating of self shows itself rooted in interrogation of and response to preconscious felt need or desire. The physiological shifting of the spinal cord–injured shows physiology redesignating the contours of physiological "self" in the sense of newly adapted mechanisms for temperature control, and new rules apportioning callous tissue and live tissue. The equilibrating physiology expresses the body's desire to continue life as a self-showing that self and other signifies. It simultaneously differentiates itself from external others and points to itself. It signifies itself and others as selves and others, both to itself and to perceiving others. The example of physiological shifting is important because it shows communication operating below and independent of the cerebral cortex. Moreover, the process of physiological shifting vigorously negotiates the boundaries between self and other, life and death that make possible one's ability to signify and display one's lived body as a symbol for oneself and others. The most significant consequence of Merleau-Ponty's view of corporeity existing as communication, however, is that communication thus reveals itself to be of and dependent on bodies and bodies of and imbedded in nature. This view effectively reconnects that which much philosophy since Plato and the rise of literacy in the West disconnects: idea and matter, spirit and nature, animals "with language" and animals "without language."

If physiological adaptation both shows life and participates in life as directed toward such self and other signification, the same might be said of technological adaptation. Walking sticks, contact lenses, hearing aids, wheelchairs, and other assistive technologies show themselves likewise of and expressing bodily needs. In coparticipation with physiological adaptive processes, technological behavior, as I show in the next section, manifests oneself as the desire to self-show—or, again, as communication.

Technological Adaptation

By "assistive technology" I mean technologies built specifically for a person with a certain disability in order to fulfill a certain task that would typically be manageable independently by persons without the disability. These extend our lived bodies beyond our physiology. The example of paralysis underlines the status of such built instruments as rooted in and expressing bodily desire.

Two key technologies for the quadriplegic are manual and electric wheelchairs and voice-activated home environment systems (thermostats, lights, computers). The felt need to be self-mobilizing and in control of one's home environment has given rise first to imagined and subsequently to built rejoinders to such needs.[25] Living with such technologies extends the lived body of the spinal cord–injured person. One's habits, health, and daily options, open up proportionately the powers afforded by the technology. Having grown accustomed to his or her wheelchair, the person's dimensions reflect those of the chair. The chair is not an object between oneself and other objects, but just as one's given body extends toward self-showing, the chair facilitates such extension. The relation between one's paralyzed body and chair is analogous to that between Merleau-Ponty's blind man and his stick (*PP* 152).

A person's lived body-and-chair requires a certain area width and length when traveling straight ahead, and a certain radial space when turning a corner. A person in a wheelchair "feels" where the chair begins and ends "without any calculation," just as a person without a chair goes "through a doorway without checking the width of the doorway against that of [his/her] body" (*PP* 143). Accordingly, rooms and options become perceivably open (or closed) to the lived body-*cum*-chair of the spinal cord–injured. The wheelchair's dimensions become in a certain sense one's own.

Assistive technology points to a self-showing that is also an expression of bodily desire. In the case of spinal cord–injured persons, the assistive technology reconfigures the trajectory of one's signifying; it allows a person to continue life as expression of given and acquired habits, habits that are or are becoming self and other, differentiating and signifying. The assistive technology both signifies human bodily desire and participates in a spinal cord–injured person's reorienting and redefining the boundaries of "self" and "other." Together with an adapting physiology, assistive technology cooperates to structure one's lived body as oriented toward a self-showing, the purpose of which is to create, mark, and sustain designation and differentiation of self and other.

If the example of physiological shifting returns communication nonreductively to physiology, the example of technological adaptation extends bodies beyond physiology and shows how one's self-demarcating self-showing necessarily adapts to include the technologically appended lived body.

Speaking the Unspeakable

Physiological and technological adaptation show themselves as two of many processes through which human corporeality signifies both itself and a beyond itself; for Merleau-Ponty, a third such process is speech. By speech I mean spoken or written language that accomplishes through words, thought.[26] The *Phenomenology of Perception* makes explicit the body as the basis of speech. Speech, like physiology and technology, shows itself as signification of and intention toward expressing bodily desire to demarcate and signify self and other.

Now, for Merleau-Ponty, traumatic experience "acquires exceptional value" because the preconscious desire to constitute the "trauma present" is enormous. It overwhelms and demands attention for itself long after the actual event has occurred. Merleau-Ponty argues that traumatic memory images, like all other memory images and behavioral processes, participate in structuring the direction of one's movements (interrogations) and perceptions (responses). Unlike "normal" memory images and processes, however, traumatic memory-images' tendency toward expression is immense. Other ordinary present experiences, by comparison, command comparatively little intention toward expression.

One reason why humans find it experientially difficult to integrate the "memory" of catastrophic experience with that of ordinary events is underlined by the etymology of the word *trauma*. Its German cognate, *Traum* (dream), points to a long-standing disjunct in human experience between integrated memory and dis-integrated memory. And, like dreams, the experiences one undergoes during a traumatizing event generally remain segmented off from conscious memory. Clinical psychologists often call such dis-integrated memory images "dissociative memory."[27]

Survivors of life-threatening events, or trauma, tend to experience the event and remember it afterward through images that are typically dissociated from ordinary memory. A traumatic experience—often life-threatening—is, says Merleau-Ponty, so unusual that the images one experiences during the event, and associates with it afterward, are placed outside the bounds of normal memory (*PP* 83). They hide out, one might say, absorbed in a condition of repression. But while these images abscond, they exert an ongoing pressure on our ordinary memory-images and daily experience. The traumatic experience

> acquires an exceptional value; it displaces the others and deprives them of their value as authentic presents. We continue to be the person who once entered on this adolescent affair, or the one who once lived in this parental universe. New perceptions, new emotions even, replace the old ones, but this process of renewal touches only the content of our experience and not its structure. Impersonal time continues its course, but personal time is arrested. Of course this fixation does not merge into [ordinary] memory; it even excludes memory. (*PP* 83)

The hiding, hard-pressing, displaced trauma-images demand attention while remaining at the same time steadfastly beyond view. One's trauma memories remain "constantly hidden behind our gaze instead of being displayed before it" (*PP* 83).

Merleau-Ponty discusses "traumatic memory," but he does not develop an analysis of trauma with respect to memory or speech. To show that the so-called representation of traumatic experience is what I would instead call "adaptive speech," and a betrayal of traumatic experience, I will critique ideas in psychiatrist Judith Herman's "Remembrance and Mourning" (*TR*).

As Herman sees it, "The fundamental premise of the psychotherapeutic work is a belief in the restorative power of *truth-telling*" (*TR* 181; emphasis added). I will show that such "truth-telling" appears to be experientially impossible, but that the activity of creating an idiom for a new narrative tale inspired by the wound is not. The created tale, insisted on by the relentless demand of the preconscious traumatic wound, fails however to represent the wound or trauma experience itself. The traumatic wound, I argue, excludes linguistic representation,[28] especially in narrative form.

Herman's own chapter implies this impossibility. The survivor is encouraged to tell the true story of her or his terror via an arduous and demanding "flooding" technique. The flooding process is called by Herman "controlled reliving of the experience," but in fact, it exposes patient and therapist to more (or, perhaps more accurately, to less) than the unsublimated trauma experience itself. Not only are the "fragmented components of frozen imagery and sensation—the smells, sounds, a pulsing heart, constricting muscles, alternating emotions of revulsion, confusion, and fear—included in this tale, but also included is a verbal retrospective, with an order and meaning attributed in hindsight" (*TR* 177). For Herman, truth-telling and storytelling engages the survivor emotionally, socially, and religio-philosophically. However, in the case of survivors who are fortunate enough to be at this second stage of Herman's three-stage recovery scheme, their emotional, social, and metaphysical orientations, are admittedly in the throes of a cocreative self-reconstruction authored by the survivor and nurtured by the therapist. Visible and significant, if unrecognized by Herman, is that the retrospective narrative of the trauma involves modes of human experience (narrative, representation, organization, meaning coherence) that were and are generally absent from the traumatic experience itself. Put differently, complex communication (language) and neural processes are generally overwhelmed and stunned during a traumatic event.[29] They are generally not able to interact with or integrate the event as usual—in part because this is no mere usual event—and this may explain why traumatic memories are often dissociated from ordinary memory (*IP*). Language is not able to be present during or for the event. In this perspective, the trauma as experienced virtually excludes language. The "true story" that Herman says demands telling and can, she says, be told, appears more precisely to be a paced eruption—into a form of complex communication—of corporeal desire driven by the power of the traumatic wound, but not representing it, or approaching a telling of "the" story, let alone "truly" so.

The forms of speech into which the wound's demand expresses itself are of the wound in only one sense—however important it may be. As Merleau-Ponty compellingly illustrates, all speech begins as preconscious desire (in this case the desire of/as the wound). For Merleau-Ponty, directedness (or desire) as such shows itself as the origin of language and speech,[30] even if it nevertheless remains absent of language and thus too, of speech. That is, bodily desire as the presupposed drive for spoken expression that has not yet become expressed, is at its root constitutively unreflected, nonverbal, and unmeaningful (*PP* 183). Speech, then, emerges of and as the bodily desire to create sense amid conditions wherein none yet is. In the case of recollecting trauma, one has impaired access to linguistic memories of the event because language was impaired during the event. This impairment together with the prereflected origin of speech is significant. It shows that the retrospective trauma narrative—as something reflected, represented and meaning-ascribed—is incidental to the traumatic experience itself. Gregg Horowitz corroborates this view in his description of uncontrolled, intrusive trauma memories (as opposed to therapeutically nurtured ones). "[W]hen the traumatic event recurs, it does so as itself, that is, as unsublimated; it thus confronts the sufferer as what cannot be represented as other than it is, which is to say, as what cannot be represented. . . . At the heart of present forms of the past's future, the past juts out; in remaining unmediated by the available forms of mediation, traumatic insistence is the ruination of the representation relation."[31]

I am suggesting that speech, understood to be an expression of its origin—pure desire—shows itself to be less a mode of representation than a mode of bodily adaptation. What I mean by "speech as bodily adaptation" can be illumined by the two parallel examples of bodily adaptation: physiological adaptation and technological adaptation. We can recall Russell's discussion of cell movements in the direction of a surface wound. The cells show the autonomic processes of anatomy moving (interrogating) and perceiving (responding to) internal and external conditions—promoting survival and self-signification. The directedness of these cells simultaneously show bodily desire constructing/negotiating its bodily boundaries and so, creatively differentiating and signifying itself as a self-designated (dynamic) whole with respect to its environs. Physiological adaptation shows animal life as directed toward such self and other signification not merely for the sake of survival, but also for the sheer making of oneself a spectacle to be seen. Use of contact lenses, prosthetics, and other assistive technologies do likewise. Both physiological and technological adaptations show the lived body actively responding to preconscious bodily impulses, and in extreme cases, aggressively negotiating bodily self-signifying survival: for instance, redrawing (in cases of traumatic bodily injury) and differentiating its lived corporeal boundaries (prosthetics, the blind man's stick, the wheelchair, etc.).

Adaptive speech, I would argue, like adaptive physiology and technology, shows itself of and expressing bodily desire and likewise the preconscious need to stabilize, demarcate, and signify oneself. It is an example of the lived body actively creating/negotiating the boundaries between self, others, and surroundings. Herman's work implies—but does not recognize—that speech in fact registers adaptation, not representation, with respect to traumatic experience. While Herman does acknowledge that horrific events destroy the self or the whole, she does not recognize that with the destruction of the preterrified self goes also any proper representation of the horror in its aftermath. While she recognizes that the ability to cope with the traumatic experience requires developing "a new self" (*TR* 198), she does not see the incongruence between the retelling of the tale by the new self and the "experience" of the trauma, itself closed off from the language processes appealed to in its aftermath.

Of specific importance, in the retelling by a new self, is that the "truth-telling" that Herman speaks of substitutes for what can be more appropriately understood as adaptive speech, which itself appears rooted in prereflective bodily desire. In the aftermath of trauma, it seems that speech—like the physiological and technological bodily processes discussed—adapts to the demands of traumatic experience by renegotiating the conditions of one's self-understanding and self-signification. This radical redrawing of self requires not only that the survivor reconnect with the new self-delineated boundaries once they are accomplished, but that she or he reconnect with others as a newly drawn self. Importantly, if one's recreated self makes use of verbal and reflective resources, it does so not because the new self is rooted in them. Like physiological and technological adaptation—which likewise manifest bodily desire as communication and self-showing as their manifest activity—the signification of adaptive speech, or in this case of traumatogenic adaptive speech, appears rooted not in the communication it shows but in corporeal desire or the intense demand of a prereflective wound.

Disclosing the source of "truth-telling" as the preconscious call of a suffering body does not render Herman's work unimportant. The disclosure is significant because it sustains Herman's ideas about trauma narrative by reorienting them through the recognition of Herman's own blind spot. By misidentifying traumatogenic-adaptive speech as a true "telling of the trauma story," Herman dismisses and remains (at least technically) in denial of the enormity and opacity of traumatic experience. She veers toward a certain "triumphalism" against which Gregg Horowitz has warned us (*SL* 130). Such triumphalism bears symptoms of denial of—or a wish to complete—one's mourning; it displays, suggests Horowitz, a symptomatic wish for the dead to stay quiet (*SL* 122).

As Horowitz's words indicate, Herman's optimism and overestimation of the power of representational speech underestimates horrific experience and relegates it to a potential but false mastery. Indeed, Herman's clinical experience if not her methods for interpreting it, shows the distance of dissociative trauma images and affects from representational form. Her interpretive choices nevertheless assume traumatogenic experience to be conformable to ordinary perceptual and reflective bounds, even when it shows itself exceeding them. The trauma as trauma, bursts out of any human habitudes and perceptual structures of reflection and linguistic representation. And so, even if a newly created narrative corresponding to the event can be told—and Herman's work suggests as much—the newly created or new trauma story remains other to the wound or traumatic experience itself. Extreme terror appears as other than human representation.

If we want to speak of keeping true to the trauma (and perhaps we do not?),[32] or of speaking in such a way that more probably attempts health, then perhaps we ought to say that traumatic experience is "told" truly as unspoken and unspeakable. And this is not to say we ought to limit ourselves to silence, but that we should not mistake the adaptive stories said in the aftermath of our dead, for more than contemporary modes of response to prereflective calls of wounds fated to remain unrepresentable. In *Hiroshima mon amour*, the film by Marguerite Duras and Alain Resnais, the character of the French actress, in dealing with the legacy of her tragically ended wartime love affair, avoids such a mistake about contemporary responses to an overwhelming wound. This nonetheless does nothing to ease her torment. She eventually gives in to her desire to tell that which she knows cannot properly be told; her sacred horror becomes desacralized, and she writhes in recognition of the horror's uncontainable excess.[33]

Conclusion

For Merleau-Ponty, one's desire to creatively delineate oneself, and self-show, continues to appear as the body's manifest activity even if in the cases of physiological or psychological trauma the "self" must entirely renew itself for this display. I introduced the notion of corporeal self-showing by articulating how for Merleau-Ponty, animal bodies appear as corporeal communication. And I emphasized that while corporeity shows itself to exist as communication, it shows itself not to be rooted in communication. On the contrary, communication—and so, too, language—emerges from corporeity. I showed that for Merleau-Ponty, animals—excluding the most simple animals—are "opened up to" their environment. They move creatively in response to signs and intend-toward the creating and negotiating of their own bodily boundaries; they intend-toward pointing to themselves and thus to others too. Like Vallier and Sjöholm, I showed that, for Merleau-Ponty, animal behavior is constitutively communication rooted in corporeity that directs itself outward and makes itself an appearance.

That corporeity exists as communication, and that such communication appears rooted in the body is significant for several reasons. It shows the body as a whole displaying activity traditionally believed dependent on the brain and cognition. In so doing it minimizes one of the problems inherited with Cartesian dualism: alienation from one's body and from nature. The assertion of a difference between mind and body tends to make humans identify the concepts of mind, idea, and immateriality with one another and to associate these ideas or properties with their understanding of self. Yet, if Descartes's legacy shrinks the distance between human self-understanding and the "divine," by attributing to the self qualities traditionally linked with the dominant European idea of God and compatible with Catholic doctrine about the soul's permanence after death, it does so for a price. Identifying with the idealist side of the Cartesian bifurcation installs as presupposition the absolute separation between one's apparent self and one's body and nature. By showing corporeity existing as a communication—but that communication is corporeity—

my reading of Merleau-Ponty "returns" communication (interrogation and response, signification, idea and thought) to bodies and bodies to nature, nonreductively.

I have magnified this exposure of selves, bodies, and nature as interpenetrating, for Merleau-Ponty, through my examples of physiological, technological, and linguistic adaptation. The example of spinal cord injury shows how physiology can respond to anatomically grave twists of fate. In the case of high-level spinal cord injury, we see the physiology adapting to its new conditions by forging new mechanisms for temperature control and new rules for apportioning calloused and live tissue—thus showing how the injured body engages in primordial forms of communication (or interrogation and response) and fights to create new borders for the physiological self. For Merleau-Ponty, this physiological shifting shows bodily desire intending toward a self-showing that self and other signifies. With respect to technological adaptation, assistive technologies such as wheelchairs, walking sticks, and prosthetics extend and adapt the lived body in ways that physiology cannot. Cooperating with physiological adaptation, technological adaptation reshapes the boundaries of one's lived body as a signifying toward also bent on self-display. Through the example of traumatogenic therapeutic narrative, speech shows itself likewise to be adaptive. That is, speech negotiates its surrounding conditions of self and roots that negotiation not in the signifying of itself it shows, but in the desire of prereflective corporeity. The basis of speech responding to traumatic experience appears to be the intense demand of a prereflective violation. Judith Herman's second stage of trauma recovery shows itself not to be as Herman indicates, a proper representation of the traumatic episode, but rather to be a linguistic adaptation of a preconscious wound that remains unacknowledged and unrepresented.

By bringing into dialogue research in spinal cord injury, trauma, and Merleau-Ponty studies, this chapter discloses a framework showing adaptive physiology, technology, and speech as interpenetrating. It provides new scientific evidence for Merleau-Ponty's view of humans and nature as coconditioning and reveals and explicates Merleau-Ponty's implied concept of trauma. This chapter's notion that corporeity exists as communication and not of communication distinguishes it from certain other critiques of modernity—namely, certain receptions of Nietzsche, Foucault, and Butler that reduce their genealogies of bodily being to cultural conditioning and cultural conditioning to ideas and ideational relativism. Moreover, by showing that for Merleau-Ponty, the body exists as communication but not of communication, my vision differentiates itself from those of G. W. F. Hegel,[34] Jacques Derrida,[35] and Ernesto Laclau,[36] which, when taken to their logical extremes, reduce perceived embodiment to language, idea, or spirit.

Notes

I am grateful to Bettina Bergo, Michael Lambek, Steve Smith, the Millsaps Works-in-Progress Group, and the Mississippi Philosophical Association for comments on this chapter at various stages. An earlier version appeared as "Nietzsche After Nietzsche: Trauma, Language, and the Writings of Merleau-Ponty," in Kristen Brown, *Nietzsche and Embodiment: Discerning Bodies and Non-dualism* (Albany: State University of New York Press, 2006), 121–49.

1. See Michael M. Patterson, "Spinal Fixation: Long-Term Alterations in Spinal Reflex Excitability with Altered or Sustained Sensory Inputs" in *Spinal Cord Plasticity: Alterations in Reflex Function*, ed. Michael M. Patterson and James W. Grau (Boston: Kluwer Academic Publishers, 2001), 79 (hereafter cited in the text as *SF*); Anton Wernig, Andras Nanassy, and Sabina Müller, "Laufband (Treadmill) Therapy in Incomplete Para- and Tetraplegia," ibid., 225–40 (hereafter cited in the text as *LT*); and Jonathan Wolpaw, "Spinal Cord Plasticity in the Acquisition of a Simple Motor Skill," ibid., 119 (hereafter cited in the text as *SPM*).

2. For the purposes of this chapter "body" and "corporeity" are synonyms and refer to any perceiving body, that is, any animal body.

3. By "communication" I mean signification. An example of primordial communication occurs in the cellular activity of all but the simplest animals when cells move (interrogate) and perceive (respond to the movement). See Maurice Merleau-Ponty, *Nature: Course Notes from the Collège de France*, trans. Robert Vallier (Evanston, IL: Northwestern University Press, 2003), 179, 219. Hereafter cited in the text as *N.* A cell's perceptive response could be said to signify to surrounding cells. What does it signify? The movement it has just now perceived. An example of complex communication is human language and its various practices—reading, writing, speaking, and thinking that depend on an alphabetic system of phonetic writing (a signifying system that signifies sounds of the spoken words signified). For Merleau-Ponty, defining language is traditionally a multifaceted task. Commentators such as James Risser, "Communication and the Prose of the World: The Question of Language in Merleau-Ponty and Gadamer" in *Merleau-Ponty in Contemporary Perspectives*, ed. Patrick Burke and Jan van der Veken (Boston: Kluwer Academic Publishers, 1993) and Duane Davis, "Reversible Subjectivity: The Problem of Transcendence and Language" in *Merleau-Ponty Vivant*, ed. M. C. Dillon (Albany: State University of New York Press, 1991) have alternately chosen not to determinately define language, allowing its varying uses to appear contextually in their chapters, or to explicitly designate its varying significations. Davis, for instance, designates "'language' to mean the physical phenomenon of a system of signifiers, and language (without single quotes) to mean the full event of human experience" (43).

4. According to Merleau-Ponty, human experience suggests that psychical and physiological functions commingle with one another almost inextricably along an "intentional arc." See *Phenomenology of Perception*, trans. Colin Smith (London: Routledge and Kegan Paul, 1962), 135–36. Hereafter cited in the text as *PP*. Merleau-Ponty's *Phénoménologie de la perception* was first published in Paris in 1945.

Terms often used to describe Merleau-Ponty's concept of intentional arc—"reaching for," "directed at," "desire for" (*PP* 55; 111; 135–36)—imply an open and only partially outlined future toward which a human being is aiming. By such language and by the term "desire" for Merleau-Ponty, I mean such indeterminate aiming.

5. Three exceptions: Janice McLane, "The Voice on the Skin: Self Mutilation and Merleau-Ponty's Theory of Language," *Hypatia* 11, no. 4 (1996): 107–18; Wendy Denton, *The Soul has Bandaged Moments: Self-Injury as a Language of Pain and Desire* (Ph.D. diss., Graduate Theological Union, Berkeley, CA, 2000); and Dan Hetherington, *Disaster Trauma: A Phenomenological-Linguistic Analysis of Buffalo Creek Flood Accounts* (Ph.D. diss., Duquesne University, Pittsburgh, PA, 2002).

6. If analyses of Merleau-Ponty's writings with respect to a concept of trauma are rare, those analyzing his texts with regard to topics somewhat related to trauma, that is, violence—Linda Martín Alcoff, "Merleau-Ponty and Feminist Theory on Experience," in *Chiasms: Merleau-Ponty's Notion of Flesh*, ed. Fred Evans and Len Lawlor (Albany: State University of New York Press, 2000); Geraldine Finn, *Why Althusser Killed His Wife: Essays on Discourse and Violence* (Atlantic Highlands, NJ: Humanities Press, 1996); John O'Neill, "Merleau-Ponty's Critique of Marxist Scientism," in *Phenomenology and Marxism*, ed. Bernhard Waldenfels (London: Routledge and Kegan-Paul,, 1984); John Somerville, "Violence, Politics and Morality," in *Philosophy and Phenomenological Research* 32 (Dec. 1971): 241–49; and Kerry H. Whiteside, "Universality and Violence: Merleau-Ponty, Malraux, and the Moral Logic of Liberalism," *Philosophy Today* 35, no. 4 (1991): 372–89—or loosely related to trauma, that is, sanity, schizophrenia or psychiatry generally, are becoming less rare—Kym MacLaren, "Emotional Disorder and the Mind-Body Problem: A Case-Study of Alexithymia," *Chiasmi International* 8 (2006): 139–55; Lucia Angelino, "Some Notes concerning the Dialogue between Merleau-Ponty and Melanie Klein," *Chiasmi International* 6 (2005): 369–81; Pierre Rodrigo, "Merleau-Ponty and Psychoanalysis: The Unconscious as 'Negative Magnitude,'" *Chiasmi International* 6 (2005): 65–85; William S. Hamrick, "Language and Abnormal Behavior: Merleau-Ponty, Hart, and Laing," *Review of Existential Psychology and Psychiatry* 18 (1982-83): 181–203; F. A. Jenner and A. J. J. De Koning, eds., *Phenomenology and Psychiatry* (London: Grune and Stratton, 1982); and David Michael Levin,

"Sanity and Myth in Affective Space: A Discussion of Merleau-Ponty," *Philosophical Forum* 14, (Winter 1982–83): 157–89.

7. Merleau-Ponty in his 1957–58 lecture course on nature (*N*).

8. This is Uexküll's position as discussed by Merleau-Ponty in his 1957–58 lecture course on nature (*N*). Examples of simple animals are the marine worm, the medusa (a small jellyfish), and the sea urchin. Merleau-Ponty distinguishes simple animals from other more complex animals to which he attributes an environment ("*Umwelt*") to which animals adapt, respond, and with which they negotiate. Simple animals by contrast, he says, are machine-like because they are virtually closed off from their surroundings; they are generally incapable of relating to, adapting to, and negotiating with their environs. To the extent that is so, suggests Merleau-Ponty, simple animals do not have an environment. In other words, to the extent they are closed off to an environment, the environment is not part of their world.

9. Robert Vallier, "The Indiscernible Joining: Structure, Signification, and Animality in Merleau-Ponty's *La nature*," *Chiasmi International* 3 (2001): 203, 205. Hereafter cited in the text as *IJ*.

10. This argument owes much of its inspiration to ideas expressed by Robert Vallier (*IJ*).

11. In addition to the explanation of Merleau-Ponty's use of the phrases "directed to" and "oriented towards" and the concept of desire given in note 4, it is helpful to be aware of the implications of the phrases and concept.

Merleau-Ponty's uses of phrases such as "oriented towards," "directed to," "desire for," and "projection at" typically are not completed by an object (*PP* 55, 111, 135–36). And yet, each includes a preposition. Traditionally, these prepositions require a prepositional object. A person "rises towards some thing." Our grammar typically assumes that an action, like "rising towards," has a determinate referent, which here is not the case.

Although for Merleau-Ponty, the concept of an intentional arc, involves a bowing, the bowing is not at a definite object or a thing. The logic structuring Merleau-Ponty's concept of intentional arc not only untethers the object from the subject, but diffuses the supposed determinacy of each. If the concept of intentional arc seems to mute or modify a subject-object reflexivity, it nonetheless amplifies one aspect of certain Western notions of self: eros or desire.

Merleau-Ponty's concept of desire, understood through his idea of intentional arc, in effect dislodges the traditional framework anchoring most sentences of European descent and the philosophical, religious, and commonsense views of self and goodness built with those sentences. "It is a question of recognizing consciousness itself as a project of the world, meant for a world [or object] which it neither embraces nor possesses, but towards which it is perpetually directed" (*PP* xvii). Our consciousness never "possesses" objects, suggests Merleau-Ponty. Our consciousness does not experience objects that can be determined by definite limits. "[S]ensations," he writes, "and images which would be the beginning and end of all knowledge never make their appearance anywhere other than within a horizon of meaning" (*PP* 15).

We can analyze the latter phrase "horizon of meaning" by concentrating on the concept of horizon. A horizon can be interpreted as an indefinite boundary. If one considers the horizon of one's field of vision right now, one will probably not see a firm line demarcating the appearing field from that beyond. Such a horizon, Merleau-Ponty suggests, is generally that which accompanies our sensation of any perceptual experience or object (*PP* 15–18). This would suggest that even if over time the meaning we ascribe to an object appears determinate, the present experience of the object to which it refers is not.

The Greek cognate of "horizon," "ὁρισμός," is illustrative here (Cf. Martin Heidegger, *Being and Time*, trans. Joan Stambaugh (Albany: State University of New York Press, 1996), 432, n. 30. The relation between the English word "horizon," and the Greek word for definition, "ὁρισμός," points the way for a helpful understanding of how definition or limit operates in the context of Merleau-Ponty's work. One can understand "ὁρισμός" to mean an indefinite limit, and Merleau-Ponty's concept of intentional arc to stretch toward such a limit. The area toward which it stretches would seem to be an imperfect "object" of subject-object mechanics. In fact, Merleau-Ponty's concept of intentional arc employs both the language of "horizon" and its etymological heritage implying experience as an indeterminacy of perceptual particulars (*PP* 22).

12. "World" for Uexküll means "objective world" as opposed to "environment," the "milieu tailored to the animal" (*N* 172).

13. Maurice Merleau-Ponty, "Themes from the Lectures," trans. J. O'Neill, *In Praise of Philosophy and Other Essays* (Evanston, IL: Northwestern University Press, 1988), 163. Cited by Robert Vallier, *IJ* 201.

14. The term "organ" used in this context is borrowed from Vallier (*IJ* 201).

15. *N* 245, quoted in *IJ* 201.

16. Although Merleau-Ponty hierarchizes orders of animals in *The Structure of Behavior*, trans. Alden L. Fisher (Pittsburgh: Duquesne University Press, 1983), by the time of his three lecture courses on nature, 1956–60, he suggests a horizontal, cocommunicating "empathy" between human bodies and animal bodies. This point is made by Vallier (*IJ* 205).

17. By "trauma" I mean either physical or psychological trauma, or both. I define psychological trauma as exposure to conditions that overwhelm one's aptitudes for having and processing experience. No single set

of conditions is in itself determinately traumatic for all humans and "[n]o two people have identical reactions, even to the same event," Judith Herman, *Trauma and Recovery: The Aftermath of Violence—From Domestic Abuse to Political Terror* (New York: Basic Books, 1997), 58. Hereafter cited in the text as *TR*. Nevertheless, "the likelihood that a person will develop a post-traumatic stress disorder depends primarily on the nature of the traumatic event" (*TR* 58). War combat and other situations that place one in an unusually vulnerable psychological and physical position have been shown to affect many people in predictable ways (*TR* 58). I define physical trauma as exposure to conditions that overwhelm one's physiological aptitudes to independently adapt to the conditions (i.e., tissue, spinal cord, brain, or organ wounds; serious illness). For evidence of relatively predictable anatomical responses to physical trauma, see Barbara L. Bullock, *Pathophysiology: Adaptation and Alterations in Function*, 4th ed. (Philadelphia: Lippencott-Raven Publishers, 1996); and Martha Freeman Somers, *Spinal Cord Injury: Functional Rehabilitation*, 2nd ed. (Upper Saddle River, NJ: Prentice Hall, 2001). Hereafter cited in the text as *PAF* and *SI*, respectively. The idea that a person's response to trauma happens "across several interpenetrating levels," braiding so-called physical and psychological planes, is mine and inspired by Merleau-Ponty's concept of perception.

18. An exception in the notes from the lecture courses (*N* 178–83) is a discussion of E. S. Russell's study of animal tissue repair from Russell's *The Directiveness of Organic Activities* (Cambridge: Cambridge University Press, 1946).

19. High-level (cervical) spinal injuries cause one to lose sweat gland function because sweat glands are controlled by the sympathetic nervous system whose base of operations—the thoracic and lumbar spinal region—is below the cervical region, and thus can no longer communicate with the brain. In the same way one loses motor and sensory functions of limbs controlled by spinal nerves located below the level of the injury, one loses other functions, such as sweat glands, if they are controlled by spinal nerves located below the level of the injury.

20. James Grau and Robin Joynes, "Pavlovian and Instrumental Conditioning within the Spinal Cord: Methodological Issues," in *Spinal Cord Plasticity: Alterations in Reflex Function*, ed. Michael M. Patterson and James W. Grau (Boston: Kluwer Academic Publishers, 2001), 13–54. Hereafter cited in the text as *PC*.

21. Patterson continues the tradition of A. M. DiGiorgio whose studies of 1929, 1943, and 1947 were among the first suggesting spinal cord memory. Patterson summarizes this tradition in terms of the work of his predecessor: "DiGiorgio had shown that a cerebellar lesion in anesthetized animals could produce an asymmetrical posture in the hind limbs, with one

flexed actively and the other extended. If the animal was left in this posture for several hours, then the spinal cord severed at the mid back, the postural asymmetry would remain. The flexion of the limb was assumed to have been 'fixated' in the spinal cord by the altered outflows from the cerebellar damage. This spinal fixation was one of the initial demonstrations of memory in the spinal cord" (*SF* 78).

22. Wolpaw shows via experiments involving the spinal stretch reflex (SSR) and corresponding electrical reflex (H-reflex), both of which are controlled by the spinal cord, that the spinal cord "plays an important part in skill acquisition" (*SPM* 101). His findings, suggesting that the spinal cord has memory and learns, challenge traditional views. Wolpaw notes that studies about "the acquisition of motor skills traditionally focus on supraspinal areas such as cerebral cortex and cerebellum" (*SPM* 101). Comparing the reflex acquisition capacity of the spinal cord to the language acquisition aptitude of the cerebellum, Wolpaw writes, "Like the changes in spinal reflexes . . . this plasticity is particularly prominent early in life (e.g., Kuhl, 1998). While languages can be learned, and spinal reflexes can be modified, later in life, this learning is clearly constrained by the patterns established in the first few years. A language acquired later on is usually spoken with an accent derived from the individual's original language, and reflex conditioning later in life probably cannot reestablish the antagonist excitation lost early in a normal childhood" (*SPM* 119).

23. Wernig, Nanassy, and Müller have shown that the spinal cords of persons chronically confined to wheelchairs due to incomplete paraplegia or quadriplegia, can acquire motor skills for walking (*LT* 225–40).

24. By "other" I intend two meanings: first, the changed circumstances, particularly the new boundaries of environment respective of the new territory of self; and second, perceivers "external" to oneself.

25. Elaine Scarry first suggested this reasoning to me in her distinction between the human process of "making up" and "making real" in response to pain. See Elaine Scarry, *The Body in Pain: The Making and Unmaking of the World* (Oxford: Oxford University Press, 1985).

26. In this chapter the term "speech" names what I call complex communication as opposed to primordial communication. A precondition of complex communication is a more intricate neural network like that of the human brain. Neural processes located below the cerebrum are comparatively less complex (*PC* 46). If for Merleau-Ponty speech and physiology are both communication—that is, forms of signification—not all communication exists as language. "Speech" for my purposes here means language. It accomplishes thought through spoken or written words.

27. For a helpful discussion of dissociation, see May Benatar, "Running Away from Sexual Abuse: Denial Revisited," *Families in Society* 76, no. 5 (1995): 318–19.

28. By "linguistic representation," I mean "speech"—that is, the interactive and integrative processes of complex communication (see note 26)—and not literal or identical representation of an object in either a Cartesian or Kantian sense. Merleau-Ponty's writings have been said to offer a "nonrepresentational" view of perception. Please note that this latter use of "representational" is different from mine. The latter is defined with respect to ontologies that presuppose a subject-object dualism (i.e., those of Descartes and Kant). According to "representation" understood in this second way—representation of a decidedly "external" world—Merleau-Ponty's writings suggest our perceptual aptitudes do not represent but rather enact interactions with the world. This reading is implied by James Risser, "Communication and the Prose of the World: The Question of Language in Merleau-Ponty and Gadamer" in *Merleau-Ponty in Contemporary Perspectives*, ed. Patrick Burke and Jan van der Veken (Boston: Kluwer Academic Publishers, 1993), 134. It aligns with the three-tiered process affording self-signification I describe. First, interrogating and negotiating an environment. Next, differentiating inner from outer. And last, presenting one's body surface as self-signifying. Such a threefold process could be said to be interactionist because it involves a reciprocal shaping across the planes of so-called subject (animal) and object (environment) thereby undermining both Cartesian dualist and Kantian rationalist representational theory.

29. Bessel A. Van der Kolk and Onno Van der Hart, "The Intrusive Past: The Flexibility of Memory and the Engraving of Trauma," in *Trauma: Explorations in Memory*, ed. Cathy Caruth (Baltimore: Johns Hopkins University Press, 1995), 172–76. Hereafter cited in the text as *IP*.

30. Desire can be said to exist for Merleau-Ponty in two respects: (1) as signification and expression; and (2) as a provisionally pure drive that is prior to and presupposed by its own subsequent expressing; thus it is presupposed by signification, language, and reflection.

31. Gregg M. Horowitz, *Sustaining Loss: Art and Mournful Life* (Stanford, CA: Stanford University Press, 2001), 124. Hereafter cited in the text as *SL*.

32. A Nietzschean critique of the value of "truth" would put into question the benefit of "keeping 'true' to the trauma," if one's motivation for doing so is solely the desire for so-called truth and not primarily a desire for life enhancement.

33. Cf. Cathy Caruth, "Literature and the Enactment of Memory (Duras, Resnais, *Hiroshima mon amour*)," in *Unclaimed Experience: Trauma, Narrative, and History* (Baltimore: Johns Hopkins University Press, 1996).

34. G. W. F. Hegel, *Phenomenology of Spirit*, trans. A. V. Miller and J. N. Findlay (Oxford: Oxford University Press, 1979).

35. Jacques Derrida, *Of Grammatology*, trans. Gayatri Chakravorty Spivak (Baltimore: Johns Hopkins University Press, 1974).

36. Ernesto Laclau, "Deconstruction, Pragmatism, Hegemony," in *Deconstruction and Pragmatism*, ed. Chantal Mouffe (London: Routledge, 1996).

Part 2

Trauma and Bodily Memory: Poetics and Neuroscience

5

Trauma and the Impossibility of Experience

Idit Dobbs-Weinstein

"Rabbi, is my wife permitted or forbidden?"

—"City of Slaughter"

Hayyim Nahman Bialik, (1873-1934) the "Hebrew national poet,"[1] gives voice to the need *or*[2] temptation to respond to a traumatic event with an escape from experience into law or some other form of ideological formal refuge. "Quoting" a husband who, while hiding, peeked between the cracks to witness his female relatives' sexual brutalization by the Cossacks, Bialik graphically puts in relief the brutality of experience and the "natural" desire to shield a wounded, suffering, ("traumatized?") psyche by *or* with the force of law. Although the "City of Slaughter" (*'Ir ha- Harega*) and its complement "On Butchery" (*'al-ha-Shchita*) vividly describe the violence performed on the bodies of some of the victims,[3] the latter remain dumb, voiceless, or silent. Although the women in the "City of Slaughter" and the butchered child in "On Butchery" were brutalized, they neither witness their violation nor can bear witness to it; rather, they are relegated to the nether regions of voiceless, nondiscursive, passive, or dead flesh. Although they may have experienced horror, their horror (if one can call it theirs) cannot be expressed; hence, one may, indeed must, ask whether they can be said to have undergone or experienced traumatic "events." Conversely, the observers of brutalization, those who theoretically and legally are expected to be able to bear witness to it, are depicted as the ones who "experience" a trauma, which experience they seek to escape by means of law *or* religion, by containing the horror within intelligible structures. Intensifying the horror, Bialik exposes the individuals' lack of significance, "victims" as well as "witnesses," by naming other "witnesses" to the violence, who alas are dumb: "the sun was shining, the wheat was blooming, and the butcher was slaughtering." In short, Bialik's poetry after the Kishiniev Pogroms, (during Easter 1903, only a bit over a century ago) turns the fear of nature (as external, inhuman, and necessarily violent)—a fear that gives rise to mythology, religion, metaphysics, and redemptive history—upside down. According to the prevailing mythology/religion *or* metaphysics, each indeed is acting in accord with the laws of its nature. The former, natural entities "perform their function" in accord with nature and are compelled to do so; the latter, are human and free to choose. Highlighting the purported opposition between the natural and the human, Bialik's poetry suggests a decisive link between the myth of violent nature and willful human violence.

There is something both profoundly insightful and repulsive about this picture—and it is this profoundly vulgar ambiguity that I seek to intensify in the following few pages (with the help of Benjamin and Adorno.) The ambiguity that I seek to uncover, rather than resolve, in this chapter, an ambiguity surrounding the terms "trauma" and "experience," let alone their "translations" into concepts,[4] is multifaceted and initially appears to result from multiple discursive origins *or* practices, literary, psychoanalytic, philosophical, legal, and so forth. But does it? May it not be the case that all these discursive practices implicitly seek to eliminate/occlude the "extremity that eludes the concept" by means of explanation or naming? To what extent is the desire/need for cure *or* intelligibility not itself a mark of a fear whose appearance is continuously deferred by multiple explanations whose purpose, ironically, is to offer hope—expiation—law, instead of the appearance (however fragmentary) of an unbearable, unknowable, and singular experience? To what extent is the insistent desire for intelligibility a desire for an escape from "vulgar" experiences, that is, singular experiences that exceed intelligibility? In short, to what extent is the desire to subdue material experience by reason an expression of an insane hope for redemption, religion, or metaphysics, whose promise not only redeems but, in fact, eliminates all singular moments of suffering? For, surely, the victims, whether dead or alive, cannot be redeemed from a "past" violence, irrespective of its "present" mode of experience, let alone, for the sake of the improvement of future generations.[5]

Just as Bialik's "poetry" balks at making sense of suffering and indicates, at least implicitly, the complicity between the desire to make sense, the narrative, *or* historical justification of slaughter and religion, almost half a century later Benjamin and Adorno extend Bialik's indictment to philosophy, especially metaphysics, and to the cultural, literary expressions that are informed by it. Although neither Benjamin nor Adorno (nor Freud, for that matter) assume that the fear and hope that give rise to religion/metaphysics can be eliminated by a materialist critique[6]—at best, a naive metaphysical assumption—their critique seeks to expose the complicity, perhaps identity, between religion/metaphysics and the redemptive history that not only justifies suffering but, more important, immunizes against, or renders impossible, its experience by means of various universalizing narratives. Benjamin's observations about the philosophy of history are still stunningly apt today:

> The current amazement that the things we are experiencing are "still" possible in the twentieth century is *not* philosophical. This amazement is not the beginning of knowledge—unless it is knowledge that the view of history which gives rise to it is untenable.[7]

> "Direct Communicability to everyone is not a criterion of truth."[8]

I began this chapter with a look at a poem that gives voice to the agony of a poet in the face of brutality and the unbearable violence it unleashes,[9] a poet who eschews both the form and content of available poetic conventions, replacing both edifying and condemnatory explanation with lengthy and vivid detail, in the manner of a chronicler, in order, first, to highlight the ambiguity of the term "trauma" and, second, to explore this ambiguity in terms of the "possibility/impossibility" of experience. Bialik's poetry is the work of Benjamin's chronicler, "who recites events without distinguishing between major and minor ones," and who thereby "acts in accordance with the following truth: nothing that has ever happened should be regarded as lost for history."[10] The poem's "failure" to distinguish between major and minor events or between nonhuman and human "witnesses," that is, between nonconscious, dumb nature and conscious "subjects,"[11] is, in fact, its strength; for it makes manifest the poet's courage to resist the temptation of making the events intelligible. In short, the poem presents but does not seek or pretend to represent.

Most contemporary discourses about trauma explicitly endorse or implicitly embrace a peculiar dualist, metaphysical dialectic between, on the one hand, a banal domestication that extends whatever may be named by "trauma" indifferently to all concrete, material forms of suffering, violence and their attendant neuroses, and, on the other hand, seeks to contain this material excess ideally, most often within teleological structures, the most prevalent and banal of which is a variant of the question of theodicy. That is, whereas most contemporary discourses seek to render trauma, even its nonavailability to symbolic representation, conceptually accessible, the poem highlights an entirely different, a-conceptual ambiguity. In the poem it is not so much trauma that resists symbolic representation but rather the subject of trauma who/that remains undecidable. Differently and succinctly stated, the poem resists the reification of the event, let alone its site, while it insists on its material singularity. In contradistinction, contemporary discourses about trauma—both those that extend it to any and every experience of violence or loss and those that seek to locate it very precisely in a dimension of psyche, often named an unconscious which is disembodied and therefore nonrepresentable—locate trauma in *or* as subjective experience, even if inaccessible to the consciousness of the traumatized subject. But who or how is the subject when, insofar as "it" cannot be represented, "its" experience cannot enter into a coherent narrative history? Insofar as the *aporiae* of both the subject of experience and the event of trauma (the "object" of experience) remain unproblematic, that is, occluded, the occlusion transforms the uniqueness of each moment of their nonconcurrence into an abstract universal; "subject" and "event" thus become peculiar metaphysical entities *or* transcendental data whose presumed facticity abrogates and denigrates the irreducible embodied singularity of traumatic events, the moment of their objective reality.

If availability to conscious representation is the decisive criterion that locates or acknowledges traumatic experiences, then either the ones who suffer trauma are not its subjects, or the objective reality of trauma, its concrete material (social as well as historical) existence can only be affirmed by the horrified witnesses (the husband, sun, and wheat of Bialik's poem), by the analyst or, distasteful as it may sound, by the perpetrator.[12] In short, and again, either the concrete, material experience of trauma cancels out all subsequent experience, annihilating the subject (and surely, in the case of pogroms, let alone Auschwitz and other genocides, this is literally true about most of the brutalized "subjects," who are nothing), or experience requires a disembodied distance, horrified by, indifferent to, or enjoying the singularity of the suffering body.[13] It is important to note, however, that to the same extent that those who are horrified by the violence, including those who "are amazed that the events we are experiencing are still possible today," seek to render the horror intelligible, so also they seek to nullify, or to become indifferent to it. In short, in every one of these instances, the place of concrete bodies and those of both "traumatic experience" and "subjects" are radically *or* substantially separate both *in dictu* and *in re*.

Now, whereas the accounts of traumatic experiences presented by practicing psychoanalysts often deploy the dualistic idiom from which I recoil, their work is rarely founded on such a dualism, nor can it be. In fact, given that the afterlife of trauma often, nay always, appears in "bodily" symptoms, the therapeutic dimension of psychoanalysis cannot disengage body and psyche, for the afterlife of trauma is always manifest as bodily symptoms.

Hence, however ill at ease I may be (and I am) with the idiom of many psychoanalytic case studies, unless they are rigidly deployed in accord with this or that theory or creed, these are not the subject of my critique. Rather, I question those appropriations of psychoanalytic language and approaches by academic disciplines, especially disciplines "in crisis," whose respectable methods and language, on the one hand, are inadequate to account for the concrete, historical experiences that inform their readings of traditional texts and, on the other hand, appear as repressive for the first time. For, ironically, the experience of history as catastrophe, an experience that undermines structures of meaning, that is, genres, schools of interpretation, and so on, almost always elicits the need/desire to replace one form of the discipline, one doctrine, one universal necessity with another. In short, the academic turn to psychoanalysis in response to history is, often, at the same time an attempt to discover a new escape from concrete history *or* a new foundation. Seeking a new foundation, a new metanarrative, where traditional ones have collapsed, especially where the complicity of the latter with concrete historical violence is neither acknowledged nor subjected to rigorous critique, often requires the insidious reinstatement of old categories

Cathy Caruth's *Unclaimed Experience: Trauma, Narratives, and History*,[14] makes manifest the manner in which psychoanalytical, literary, and philosophical critiques that interrogate canons without exhibiting their optionality, that is, lack of necessity, are inextricably bound by their uninterrogated categories that, as such, become insidious. One example must suffice to substantiate my claim. In her introductory description of trauma (trauma that would later be understood both individually and historically), Caruth presents what she takes to be Freud's understanding of the enigma of trauma as follows:

> [T]he term *trauma* is understood as a wound inflicted ***not upon the body but upon the mind***." But, what seems to be suggested by Freud in *Beyond the Pleasure Principle* is that the ***wound of the mind***—the breach in the mind's experience of time, ***self***, and the ***world***—is not, ***like the wound of the body***, a simple and healable event.[15]

Now, it is not merely the use of dualist language in Caruth's description of trauma that I find problematic but rather and, above all, the fact that the language implicitly or explicitly reflects a thoroughly Cartesian real (metaphysical) separation between body and mind. Moreover, and even if I overlook the questionable attribution of mind-body dualism to Freud, Caruth's distinction between self and world entails that the mind is the subject-self whose wound renders it incapable of representing itself to itself and hence healing itself as an internal claimed "objective experience."

In contrast, the body is merely an external object (an extended, three-dimensional mere object of sense) whose wound is simple and healable. It is not surprising, therefore, that from this metaphysical dualism Caruth draws an epistemological conclusion that entails that all bodily events/experiences are fully representable and "knowable," whereas trauma marks a separate mental experience that, insofar as it cannot be represented, cannot be known. In short, for Caruth, there are mental experiences that are absolutely independent from concrete, material embodiment.

Rather than Caruth's (implicit?) assumption that some experiences do not depend on the singular, materially, and historically constituted body, rather than engage in a critique of Caruth's idiosyncratic reading of trauma and history, I wish to turn to an alternative, radically a-dualist reflection on trauma, experience, and history.

Since the ambiguity intrinsic to "trauma" has become progressively inaccessible to critical analysis, owing to the proliferation of discourses that seek to cure it, that is, eradicate its horrifying and singular reality, or its irreducible singularity, by means of intelligibility or meaningful discourse, and since this ambiguity is amply evident or at least is rendered visible through significant disagreements among the chapters that address trauma directly in this volume, I shall confine the rest of my remarks to the problem of experience and history or, more precisely, to the history that simultaneously renders experience impossible and covers over this impossibility, a history that is ahistorical, and whose origin is religion *or* metaphysics, a history that renders spectators and perpetrators of violence complicit in the same violence and leaves the victims, dead and alive, responsible for their violation.

> All too often the presupposition is that anti-Semitism in some essential way involves the Jews and could be countered through concrete experience with Jews, whereas the genuine anti-Semite is defined far more by his incapacity for any experience whatsoever, by his unresponsiveness.[16]

Whereas the third part of *Negative Dialectics,* the work best known and least palatable to American philosophers, is addressed to philosophers and constitutes a concrete, material indictment of philosophy's immanent and a-critical complicity with repressive institutions and ideologies; whereas it takes the form of violent, fragmentary confrontations between philosophy's abstract claims and the "vulgar" experience that philosophy, especially Kantian and post-Kantian philosophy, disdains, a disdain that renders possible unthinkable brutalities, Adorno's more popular essays, first delivered as radio addresses, especially, "The Meaning of Working Through the Past" and "Education after Auschwitz,"[17] initially seem to offer a linear didactic narrative, and to reflect a naive, hopeful belief in the power of reason to prevent the recurrence of unthinkable brutalities. These, then, may, perhaps, be read as "an emanation of the insane wish of a man killed twenty years earlier."[18] It is true that Adorno, unlike Horkheimer, never resigned himself to philosophy's or, more precisely, critical thinking's total impotence against barbarism, but it is also true that by philosophy he intended no more than a relentless and continuous immanent critique of political institutions and practices that, inter alia, lays bare their material origins and thereby exposes their contingency.[19] Insane or impossible as it may be, Adorno's wish or hallucinatory dream is that Auschwitz should never happen again. Broadly stated, the question to which he returns continuously, a question that in fact predates the horrors of Auschwitz, is, "What are the concrete, material conditions immanent in 'highly civilized society' that produce barbarism?" More specifically, especially in the later works, the question is, "What are the concrete, material conditions that render highly civilized, 'perfectly rational' human beings incapable of experience?"

Before I turn to an all too brief discussion of Adorno's very subtle and elusive understanding of experience, and the overwhelming obstacles to it, I wish to quote at some length the paradoxical manner in which he formulates the only imperative that can justify philosophy today in "Education After Auschwitz," an imperative that both sheds light on what he may mean by "experience" and ironically requires the capacity for experience in order to be truly heard.

> The premier demand upon all education is that Auschwitz not happen again. Its priority before any other requirements is such that I believe *I need not and should not justify it.* I cannot *understand* why it has been given so little concern until now. *To justify it would be monstrous* in the face of the monstrosity that took place. Yet, the fact that one is so barely conscious of this demand and the questions it raises shows that the monstrosity has not penetrated people's minds deeply, itself a symptom of the continuing potential for its recurrence as far as peoples' *conscious and unconscious* is concerned. Every debate about the ideals of education is trivial and inconsequential compared to this single ideal: never again Auschwitz. It *was* the barbarism all education strives against. One speaks of the threat of a relapse into barbarism. But it is not a threat—Auschwitz *was* this relapse, and barbarism continues as long as the fundamental conditions that favor this relapse continue largely unchanged. That is the whole horror. The societal pressure still bears down, although the danger remains invisible nowadays. It drives people toward the unspeakable, which culminated on a world-historical scale in Auschwitz.[20]

Insofar as Adorno's task is to justify the demand that should not be justified, and insofar as the need for justification is unintelligible, it is an impossible, monstrous task, the expression of a striving not so much to eradicate the conditions for the recurrence of Auschwitz but to interrupt its current occurrence, or to quote Benjamin "to blast open the continuum of history."[21] For, insofar as and to the same extent that Auschwitz is relegated to the past, so does its persistence remain invisible and inaccessible to experience. Or in Benjamin's quasi-prophetic idiom twenty-six years prior to Adorno's concern with "vulgar experience": "every image of the past that is not recognized by the present as one of its concerns threatens to disappear irretrievably,"[22] to be lost to memory and thereby to experience. Whatever Adorno may mean by experience, then, depends on an understanding of history as present memory fragments of singular events that cannot be subsumed by, in fact defy, universal, that is, redemptive, history. And, it is Adorno's claim, a claim with which I fully concur, that the possibility of all human experience today, depends on the experience of Auschwitz. Thus, the belief that Auschwitz is a past event, an event that either has been or could be worked through so as to overcome or forget it, so as to "remove it from memory,"[23] and get on with the "business of life," is a symptom of its present persistence in a form that is even more monstrous because it is even more insidious, shielding the current forms of Auschwitz from view.

Bluntly stated, Auschwitz is not an aberration but the result or effect of a progressive view of history, or Weltanschauung that uncritically accepts the necessity of historical events, a view that both produces a peculiar historical stoicism and occludes the optionality, that is, the materiality of social and political repressive institutions and thereby renders them immune to critique. But the causes of repression are to be found in concrete material conditions and institutions rather than in abstract historical constructions. History is indeed the history of the victors but the acquiescence to its necessity, to the present either as a mode of overcoming the past or as the mere passage from past to future, leads to the destruction of memory and with it the capacity for experience, leaving repressive social structures and the psyches they generate unchanged. It cannot be overemphasized that, with respect to memory and the capacity for experience, there is no difference between the understanding of the present as "overcoming" the past or its "passage" to the future; for, in both instances the present is pure facticity, fully present and fully representable within intelligible structures. While it is true that the triumph of Spirit or rationality is the triumph of brutality, and "that civilization *produces* anticivilization and increasingly *reinforces* it,"[24] unless a different story can be told, unless a different relation to the past forged, "even the dead will not be safe from the enemy if he wins," and as Benjamin reminds us, "this enemy has not ceased to be victorious,"[25] a victory which, according to Adorno, ensures that "the murdered are to be cheated out of the single remaining thing that our powerlessness can offer them: remembrance."[26] The facticity of the past as the overcoming of the present as fully representable, of the present as present to consciousness, obliterates the meaninglessness *or* unintelligibility of trauma, and with it the capacity to remember, to experience the past as a material gap of a wounded psyche. Thus understood, memory as the storehouse of/for consciousness is nothing but the substantial capacity of a subject/self that exists independently of its concrete, material experience and hence, quite literally, cannot remember.

Now, neither Benjamin, nor Adorno, nor the "angel of history" naively, mystically, romantically, or idealistically assume that any power can redeem the past or awaken the dead. They can, however, offer powerfully critical analyses *or* genealogies of the present, disclosing the manner in which "there is no document of civilization which is not at the same time a document of barbarism."[27] Rather than offer "empty" hope, their critiques seek to make manifest that and how the insistence on the pastness of the past, on the inevitability of the present and the hopefulness of future redemption/meaningfulness, is the mark of the historicism that is itself the product of an ideology/mythology/religion that "empathizes" or seeks to identify with the victor, a mythology whose destruction is the concrete condition for the possibility of experience, and hence the task of the historical materialist—the impossible and nonetheless only task still worthy of philosophy.[28]

Although, properly speaking, the victors or objective conditions of violence today are the totalizing institutions of the administered world and the pervasive ideology that intends to crush all "aberrant" thinking and conduct, (i.e., individual resistance to totalization), under the guise of democratic, that is, universal culture, Adorno explicitly questions the possibility or power of thinking to overcome objective conditions. As stated earlier, the denial that Auschwitz persists today, is a symptom of its most insidious form. Ironically, it is precisely because the demand placed on thinking today is also a demand to resist the totalizing concepts that were once paraded as the triumphant achievement of enlightened, liberating philosophy, precisely because totalizing concepts either are or can lend themselves to indefinite modes of ideological appropriations that relentlessly seek to undermine the capacity for self-reflection, the "subjective" condition for experience, that no critique of ideology or society can change the objective conditions.

Now, although Adorno's turn to the subject may seem to be a turn away from the objective, that is, material conditions, and hence to betray or at least suspend his commitment to his earlier dialectical, historical materialism, I would like to propose that it is rather a turn from the macro to the micro material conditions, from world historical explanatory systems that justify the existing oppressive institutions, to the concrete material *or* "vulgar" experiences that belie such systematic explanations. For, insofar as Adorno's turn to the subject is the turn to the personality types of the individuals who compose the collective psyche—individuals who either acquiesce to the violence, sanction it, or enjoy it and whose collective coalescence is prerequisite to the continuous success of barbarism—rather than a turn to the (imperious) fully conscious metaphysical subject, Adorno turns toward the empirical subject, the only subject whose capacity and incapacity for experience can provide the concrete material still accessible to critique. Narrowing the "subject" of critique further, and in light of the insight that real knowledge of the "positive" qualities of the victims is not simply useless but impossible when the perpetrators are incapable of experience, Adorno focuses strictly on the psychology of the perpetrators of violence. Adorno is quite emphatic that the turn to the subject is not a turn away from "objective" or concrete conditions:

> I wish, however, to emphasize that the recurrence or non recurrence of fascism in its decisive aspect is not a question of psychology, but of society. I speak so much of the psychological only because the other, more essential aspects lie so far out of reach of the influence of education, if not of intervention of individuals altogether.[29]

Describing some of the conditions and mechanisms that produce the personality type that he reluctantly names the "reified consciousness," a consciousness that, having assimilated itself to things, assimilates others to things so that it turns human beings into specimens or an "amorphous mass" that can be "finished off," [or be named today the "collateral damage" of "smart bombs"], Adorno explicitly identifies this consciousness with a blindness to history. "Above all, this is a consciousness blinded to all insight into its own conditionedness, and posits as absolute what exists contingently."[30] Denying its own conditionality, that is, materiality, and historical situatedness the blindness at the core of the reified consciousness is accompanied by a profound coldness that renders it utterly indifferent to suffering. Viewed from the perspective of its relation to history, the reified consciousness is a monstrous caricature of absolute spirit, whose necessary progress, on the one hand, cancels out every particular, let alone singular event and, on the other hand, justifies it.

Before proceeding, it cannot be overemphasized that the fascism which, for Adorno, is a question of society and beyond the "influence of education, if not the intervention of individuals altogether," the fascism that is embodied in social-economic-political institutions that produce the "reified consciousness" affects "victim," "perpetrator," and "witness" alike, albeit in different modes. Moreover, the difference between "victim," "perpetrator," and "witness" cannot be reduced to the psychology of individuals. Rather, I wish to insist, again, that no analysis of suffering, let alone "trauma," can proceed only at the level of the individual psyche; for, the individual psyche is individuated by means of social-economic-political institutions, including the educational ones. Moreover, since individual psyches, qua individual, neither precede nor succeed their institutional formation, they cannot serve as models for an interpolation of a historical narrative of "psyches and their destinies" as they do for Caruth.[31] Rather, with Adorno, I wish to turn for a brief moment to Adorno's reflections on Freud's *Group Psychology and the Analysis of the Ego* and the insights it provides to the "social-psychological relevance of talk about an unmastered past," especially at historical moments when "collective identification breaks apart."[32]

The "reified consciousness," or "authoritarian personality," which is the concern of Adorno's later essays, especially "The Meaning of Working through the Past" and "Education After Auschwitz," is one whose incapacity for experience is produced by existing repressive institutions/ideology/propaganda, and whose existence consists of nothing other than the identification with "power per se, prior to any particular content."[33] There is no "I" prior to the identification with the collective, no experience or memory independent of it and, certainly, no history apart from it. Where no panic accompanies the apparent breakdown of collective identification, Adorno concludes that the breakdown is no dissolution. Again, like a monstrous specter of the Hegelian Spirit, nothing is lost to the totality, it and only it persists.

> If the lessons of the great psychologist are not to be cast to the wind then there remains only one conclusion: that secretly, smoldering unconsciously and therefore all the more powerfully, these identifications and the collective narcissism were not destroyed at all, but continue to exist.[34]

Auschwitz persists. Insofar as, and to the extent that, its persistence is denied, insofar as, and to the extent that, it is understood to be a past event, so it is preserved, is insidiously protected, and is all the more powerful.

But, if it is indeed true that Auschwitz not only persists today but prevails, and I am convinced that it does, how can the capacity for experience be retrieved without an appeal to any normative ideal, apart from the paradoxical "Auschwitz should never happen again" while it is happening? Apart from a continuous appeal to education as the only possibility for the emergence of critical thinking, Adorno takes a rather strange and surprisingly Spinozist turn. It cannot be overemphasized, however, that the appeal to education is an exemplary instance of the impossible/possibility that constitutes the task of philosophy after Auschwitz, an appeal whose success is highly dubious in light of the abysmal "discussion" with educators following Adorno's lecture.[35] Against the common philosophical/ideological, quasi-Aristotelian conviction that society is based on appeal or attraction, or some other humanist conviction, he argues that it is based on "the pursuit of one's own interest against the interests of everyone else."[36] While this claim may sound more Hobbesian than Spinozist, and while it may appear to legitimate, rather than undermine the Leviathan, Adorno's deployment of "self-interest" is in the spirit of Spinoza rather than Hobbes.[37]

Precisely insofar as the striving for self-preservation is a striving for what is passionately believed to be one's advantage, precisely because it is the same striving, whether informed by reason or passion, for reason too is desirative, and precisely because only a passion can limit or overpower another passion, education may, and this is a weak "may," be able to make evident the self-destructive nature of prevalent beliefs.

Given the limitation of time, allow me to end this chapter where Adorno ends "Education After Auschwitz." Recalling a question by Benjamin of "whether there were really enough torturers back in [Germany] to carry out the orders of the Nazis," Adorno responded that, indeed, there were. Rather than understand the question as naive, Adorno points out that

> it has a profound legitimacy. Benjamin sensed that the people who *do* it, as opposed to the bureaucratic desktop murderers and ideologues, operate contrary to their own immediate interests, are murderers of themselves while they murder others. I fear that measures of even such an elaborate education will hardly hinder the renewed growth of desktop murderers. But that there are people who do it down below, indeed as servants, through which they perpetuate their own servitude and degrade themselves . . . against this, however, education and enlightenment can still manage a little something.[38]

Would Adorno still believe that education today "can still manage a little something?" I doubt it, unless his address to the educators in "The Meaning of Working through the Past" and "Education After Auschwitz" is understood to be an appeal to their self-interests or conatus for self preservation, rather than an appeal to reason, the originary forms of which conatus are hope and fear, the only "vulgar experience" of the empirical subject that cannot be eliminated without extinguishing its life.

Notes

1. When I was growing up in Israel the name Hayyim Nahman Bialik functioned both as a proper name and as an analytic definition. It also functioned as a symbol. Since Bialik's poetry is not the subject matter of this chapter I cannot comment on the proliferation of meanings associated with his name, although it is certainly worth noting. Thus, to say Bialik was to say the Hebrew national poet. In this chapter all references to Bialik's poetry will be based on recollection or embodied memory. While Bialik's

poetry is fully available to me both in Hebrew and in translation, I shall deliberately "rely" on memory for reasons that will become evident at the end of the chapter. For, this memory has determined my capacity for experience as well as its early limits. I wish to note that the transliteration of Hebrew words will (unfortunately) depart from "proper" forms since I cannot expect the editors of this book, nor the publisher of this series, to be able to accommodate Semitic transliteration. Thus, for example, instead of an accented "s," I use "sh."

2. I wish to note that, henceforth, every instance of the use of an italicized *or* should be read as an inclusive "and/or."

3. I wish to note that the deployment of terms such as "victim," "suffering," and "witness" seeks to underline their normative ambiguity rather than accept their conceptual force.

4. Throughout the chapter, by "concept" I intend precisely the universal and self-identical metaphysical "entities" *or* "powerful fictions" that are the mainstay of modern philosophy. I use the expression "term/concept" where I am convinced that the term has been reified and has become a universal signifier.

5. The distinction between two modes of experience alludes to Kant's political writings, especially to his explicit contempt for the appeal to "vulgar experience" as a source of knowledge, including historical knowledge. Lest it be claimed that Kant's writings on history are precritical (as my friend and former colleague Jay Bernstein claims), I offer an example taken from the *Critique of Pure Reason* (A 316–17; B 373–74): "A constitution allowing the *greatest possible human freedom* in accordance with laws which ensure *that the freedom of each can co-exist with the freedom of all others*, . . . is at all events a necessary idea. . . . It requires that we abstract at the outset from present hindrances. . . . For there is nothing more harmful, or more unworthy of a philosopher, than the *vulgar* appeal to an allegedly contrary experience, which would not have existed at all if the above measures had been taken at the right time *in accordance with ideas*, and if *crude* concepts, for the very reason that they derived from experience, had not instead vitiated against *every good intention*" (emphasis in the last sentence added). In his polemic against Mendelssohn, in "Theory and Practice," Kant is even more explicit: "It is quite irrelevant whether any empirical evidence suggest that these plans, *which are founded only on hope*, may be unsuccessful." See *Political Writings*, trans. Hans Reiss, 2nd ed. (Cambridge: Cambridge University Press, 1991) 89. It cannot be overemphasized that Kant's continued insistence on hope and the promise of human progress refuses precisely the "crude" or "vulgar" experiences of human barbarism as the evidence that belies the idea of progress, teleology, purpose.

6. Henceforth, I shall use "religion/metaphysics" to signify their occluded identity.

7. Walter Benjamin, "Theses on the Philosophy of History," *Illuminations*, trans. Harry Zohn (New York: Schocken Books, 1969). "Thesis VIII," 257. Henceforth cited in the text as "Theses." German: "Über den Begriff der Geschichte," *Illuminationen (Ausgewählte Schriften 1)* (Frankfurt am Main: Suhrkamp Verlag, 1977), 255. There are two distinct German terms that are translated by the English term "experience," namely, "*Erfahrung*" and "*Erlebnis.*" Notwithstanding their differences, as Howard Caygill argues, "[a]ll of Benjamin's writings, whether dedicated to literature, art history, or the study of urban culture, may be read as an anticipation of a 'coming philosophy'. At the heart of this new philosophy is a *radical transformation* of the concept of experience bequeathed by Kant's critical philosophy." See Howard Caygill, *The Colour of Experience* (London: Routledge, 1998), 1. Cf. Richard Wolin, Experience and Materialism in Benjamin's *Passagenwerk,*" in *Benjamin: Philosophy, Aesthetics, History*, ed. Gary Smith (Chicago: University of Chicago Press, 1990), 210–27. Irrespective of differences among the interpretations of "experience" in Benjamin's work, let alone between Benjamin's and Adorno's deployment of the term, since my concern in the current chapter is not intended as an interpretation of Benjamin's and Adorno's work, I shall forego a discussion of these differences and shall deploy the term "experience" in the sense given to it in Adorno's "The Meaning of Working through the Past" and "Education After Auschwitz," in *Critical Models: Interpretations and Catchwords*, trans. Henry W. Pickford (New York: Columbia University Press, 1998), 89–103 and 191–204, respectively. Henceforth cited in the text as *CM.*

8. Theodor Adorno, *Negative Dialectics*, trans. E. B. Ashton (New York: Continuum, Seabury Press, 1979), 41. Henceforth cited in the text as *ND.* "Kriterium des Wahren ist nicht seine unmittelbare Kommunizierbarkeit an Jedermann," *Negative Dialektik, Gesammelte Schriften, Vol. 6* (Frankfurt am Main: Suhrkamp Verlag, 1973), 51.

9. As will become evident, the "witnesses'" desire for explanation, for legal expiation, is itself a violent effect of brutality. The *arche* of brutality is that of the "desktop murderers"; they are indeed its architects, the violent execution is carried out by underlings, who, according to Adorno, are murderers of themselves, as will become evident later. The architects of brutality are grotesque specters of Aristotle's statesmen. See *The Nicomachean Ethics*, 1094a25–1094b11.

10. *Illuminations*, Thesis 3, 254. (German: 252)

11. The "subject" that this chapter seeks to put into question is precisely the post-Cartesian "I," who is the "subject of consciousness" and who is expressly defined in opposition to the "object of consciousness."

12. This is clearly Claude Lanzmann's view both in his films, best known of which is *Shoah*, and in "The Obscenity of Understanding: An Evening with Claude Lanzmann," in *Trauma: Exploration in Memory*, ed.

Cathy Caruth (Baltimore: John Hopkins University Press, 1995), which is an account of an abysmal encounter between Yale psychoanalysts and academics interested in psychoanalysis, and Lanzmann. Lanzmann's attempted conversation with psychoanalysts is remarkably similar to Adorno's attempted discussion with educators, following his lecture "The Meaning of Working through the Past," as will become evident later. Less blatantly and distastefully, the witness of perpetrators is an important dimension of Adorno's "turn to the subject" as the only available possible resistance to barbarism.

13. Most recently, recall the vivid photographs of widely grinning American soldier/"liberators" at Abu Ghraib.

14. Cathy Caruth, *Unclaimed Experience: Trauma, Narratives, and History* (Baltimore: John Hopkins University Press, 1996). Henceforth cited in the text as *UE.*

15. *UE* 3–4, bold emphasis mine.

16. *CM* 101 and ff. Cf. appendix 2, 307–8.

17. *CM* 90–103, 191–204, respectively. See also appendix 1, 295–306.

18. *ND*, "After Auschwitz," 363.

19. Cf. *CM*, "Resignation," 289–93.

20. *CM*, "Education After Auschwitz," 191. It is worth noting that this essay was first delivered as a radio address to German educators on April 18, 1966. Emphasis added.

21. "Theses," Thesis XVI. On the impossible possibility that is the "task" of thinking after Auschwitz, the demand placed on thinking today so as to resist both resignation to, and complicity with, barbarism, cf. *CM*, "Resignation," 289–93; *ND*, pt. 3, "After Auschwitz," 361–65; and Theodor Adorno, *Minima Moralia: Reflections from Damaged Life*, trans. E. F. N. Jephcott (London: Verso, 1991), pt. 3, 152 "Finale," 247.

22. "Theses," Thesis V. All of Benjamin's writing, and most prominently the "Theses of the Philosophy of History," are interruptions. This is also how Adorno understands them quoting Benjamin: "In *One-Way Street* he wrote that citations from his works were like highwaymen, who suddenly descend on the reader to rob him of his convictions." From Adorno's "A Portrait of Walter Benjamin," in *Prisms*, trans. Samuel and Shirley Weber (Cambridge, MA: MIT Press, 1990), 239.

23. *CM*, "The Meaning of Working through the Past," 89.

24. *CM*, "Education After Auschwitz," 191. It is important to note that Adorno sees this insight into civilization as one of Freud's great insights.

25. "Theses," Thesis VI, 255.

26. *CM*, "The Meaning of Working through the Past," 91.

27. "Theses," Thesis VII, 256.

28. In his relentless critique of any dogmatic form of historical materialism in the "Theses" Benjamin is at pains to distinguish his thinking from

crude, utopian, or positivist historical materialists (just as Marx attempted to do before him, albeit unsuccessfully). Thus, if later critiques of historical materialism include Benjamin's "Theses" in the category of historical materialism, they clearly misread him and, perhaps, Marx. Differently stated, what Benjamin and Adorno understand by historical materialism cannot be assimilated to a doctrine, universal method, or system, and is decidedly nonteleological. A discussion of this radical difference is clearly beyond the limits of this chapter. Suffice it to insist that Benjamin's "Theses on the Philosophy of History" and Adorno's *Negative Dialectics*, as well as "A Portrait of Walter Benjamin," make this amply evident. For both Benjamin and Adorno, the "materiality" of historical materialism is decidedly not metaphysical or ontological, on the contrary.

29. *CM*, "Education After Auschwitz,"194.

30. *CM*, "Education After Auschwitz," 200.

31. Since this chapter is not intended as a critique of Caruth, I am withholding any and all genuine critique of her work, especially *Unclaimed Experience,* which I have cited. It is important to note, however, that not only does Caruth interpolate from individual psyches to history, but that she draws no distinction between concrete individual psyches and mythological ones. Suffice it to point out that all of Adorno's work is devoted to a critique of such mythology, which critique is heir to Spinoza's critique of prejudice/religion/metaphysics. One wonders, therefore, what Caruth may mean by history.

32. *CM*, "The Meaning of Working through the Past," 96 and ff.

33. *CM*, "The Meaning of Working through the Past," 94.

34. *CM*, "The Meaning of Working through the Past," 96

35. *CM*, Appendix 1, "Discussion of Professor Adorno's Lecture 'The Meaning of Working through the Past,'" 295–306.

36. *CM*, "Education After Auschwitz," 201.

37. Baruch (Benedict) Spinoza, *The Collected Works of Spinoza: Ethics*, trans. Edwin Curley (Princeton, NJ: Princeton University Press, 1985): "Each thing, as far as it can by its own power, strives to preserve itself in its being" *E* 2P6, 498. I cite one example, but it cannot be overemphasized that the *conatus* for self-preservation forms the core of Spinoza's philosophical works and its use is prevalent in all of them. On Spinoza's materialist, political heritage found, inter alia, in the thought of Marx, Freud, Benjamin, and Adorno, see Idit Dobbs-Weinstein, "The Power of Prejudice and the Force of Law: Spinoza's Critique of Religion and Its Heirs," *Epoché* 7 no. 1 (Fall 2001): 51–70; and "Whose History? Spinoza's Critique of Religion as an Other Modernity," *Idealistic Studies* 33 nos. 2–3 (Summer–Fall 2003): 219–35.

38. *CM*, "Education After Auschwitz," 203–4.

6

Trauma's Presentation

Charles E. Scott

The secret source of humor itself is not joy but sorrow.

—Mark Twain, *Puddin'head Wilson's New Calendar, 1897*

It's a question of how trauma happens and how it carries on. There's distance to it as it comes presented. It can be presented to us photographically or verbally, more or less dramatically. Often without words or images. It might be presented in pictures or in a twisted smile on the palsied face of a child who was thrown by his mother against a wall when he was two months old—presented in a paralyzed face that shows while smiling a trauma that is past, certainly, and also strangely present. Traumas persist—remain as presented—in symptoms that carry forward the shock, in manners of sorrow and mourning, in phobias, obsessions, self-pity. These are all manners of memory in which a terrible shock or injury is infused into lives later and after the initiating event. It is infused as well by a dimension of detachment and indifference, a dimension that comes with the differentiating forces of what we call trauma. This dimension is the subject of these remarks.

Seeing Traumatic Events from Afar

Susan Sontag writes of "photography's view of devastation and death" in a 2002 issue of *The New Yorker* ("Looking at War," December 9). Looking at pictures of corpses, of Vietnamese children screaming, burning, running from their village, doused with napalm, of ribbed children starving—what are we doing? We are at the very least seeing what we believe to be an accurate portrayal of instances of human devastation, seeing, Sontag points out, by means of a recording machine, a camera, that in spite of its inevitable perspectives gives us a sense of objectivity that the art of Goya, for instance, cannot give. Sontag shows that while early photos of disasters were often staged, during the last half century such counterfeit practices have been largely abandoned. We have an increase of unstaged and untampered presentations in part because of the extraordinary technical capacity of cameras to go where the action is and to record it as it happens and in part because the near omnipresence of television's immediacy makes photographic forgery very hard to pass unnoticed. A recording device in human hands bears witness to traumatic events with absolute indifference to the scene, while the differentiating effects through the reproductive and defused perspectives of news media and commentators can carry a huge impact and make important differences in attitudes and policies.

In her observations on "the iconography of suffering," Sontag calls attention to that segment in the history of art that presents "hard-to-look-at cruelties from classical antiquity—the pagan myths, even more the Christian stories [that] offer something for every taste. No moral charge attaches to the representation of these cruelties. Just the provocation: can you look at this? There is the satisfaction of being able to look at the image without flinching. There is the pleasure of flinching" (88). Another dimension of indifference emerges: not that of the machine or of simple curiosity but that of an act of seeing and digesting what is seen. It is an indifference of distance—it's awful but it's not my face being shot away or anyone's in my immediate proximity. Seeing the picture, I am entirely free to look carefully, to see the engraved image of the bursting nose just as it breaks away from the cartilage that formed it and see the red exploding grains begin their trajectory of dissolution through the air. Gosh, I wonder what that instant feels like. This is so terrible! War is evil! "The rest of us [we who see the photographs] are voyeurs, whether we like it or not" (89), she says. There is an indifference in the act of perceiving to which a voyeuristic viewing is entirely attuned.

There's an indifference of presentation built into our seeing even when we are also outraged by what we see. It's the availability of trauma for nontraumatic, perceptive experience that I am focusing on at the moment, the availability that defines a space of nontrauma and carries with it simple unconcern for what is experienced. Regardless of our manner of response—horror, Schadenfreude, curiosity, moral outrage—the event comes as a presentation, more or less reliably fashioned in its particulars and frankness, providing a framed and limited objectivity, a report (no matter the manner in which we receive the news). Such presentations of trauma and their perception in books, newspapers, and television are themselves not necessarily traumatic, and they produce a situation of extreme distance from what nonetheless appears with greater immediacy than is otherwise available to most people. Pitiless delectation seems to be an important part of making traumatic horrors come, as it were, to life for those at a distance from them—and, as we shall see, for those who have suffered them.

Would we want it otherwise? Would we want traumas portrayed traumatically? Would we want nonvirtual trauma, real trauma in the presentation? Probably not. Better to let trauma be gone. The indifference of this distance seems to have considerable value, a value worth preserving, a value best not moralized and criticized while we benefit from it. It's the value of letting trauma go even in presenting it.

Trauma's Body

A trauma is not necessarily at first a meaningful event, but it always makes a physical difference. It produces emotions. A trauma is a neurological recording no matter whether it comes in the form of a realization that a previous and relatively painless experience was one of sexual abuse or an immediate experience of pain and violation. Some traumas do not arise from events that are directly harmful to a body. They can arise from the force of values and meanings that tell us about what happened to us, tell us that because it happened we are not who we think we are, tell us indelibly, perhaps, that we are going to die, that we are terribly vulnerable in our lives, that we have unwittingly done an evil thing, or that we are victims. These are traumas of identity. We can polish them and keep them on display because they arouse sympathy or pay dividends or make us feel important. Or they can provide a painful basis for new discovery, for the formation of new values, for growth in our sense of ourselves. Traumas have meaning for good or ill in social worlds and contexts of words, images, truths, and personal histories. But there is also a dimension of traumatic happening that is neither limited nor exhausted by meanings and values.

As we consider this physical dimension, let's assume that we are in a dimensional and complex region, that the dimensions of meaning and those without meaning happen together for most people most of the time. I am not moving toward a conclusion that separates meaning and nonmeaning in human experience and that would say in effect that people have experiences with no meaning at all. But that we are able to perceive and describe dimensions of experiences that are without meaning, and the loss of a sense for these dimensions (as well as for a dimension of indifference) is detrimental for our attunement to our lives.

Trauma happens as a somatic disturbance.[1] The regulation of a person's emotional responses and survival behavior (which is located in the cerebral cortex and in the brain's limbic system) are impacted extremely in a traumatic event. A series of instinctual processes take place as a body registers an impact that is felt as life-threatening: the limbic system signals cells to prepare for drastic action; the autonomic nervous system becomes dominant and sends hormonal signals by way of the amygdala, the hypothalamus, and the adrenal glands to all crucial organs, flooding the bloodstream with special chemicals and hyperactivating neurotransmitters. Respiration and heart rate increase and provide more oxygen for muscles. Blood is sent away from the skin so the muscles will have more of it. The body moves instantly away from homeostasis and prepares to protect itself, whether by a freezing response (tonic immobility) or by increased movement. These are all automatic survival actions with physical lineages that vastly exceed both the stretch and capability of human consideration. We may call this a normal and healthy adaptive survival response. At best it subsides when the impacting danger is past.

This kind of somatic occurrence, however, does more than respond to dangerous immediacies. It produces a prereflective memory trace that can operate as though the past danger were present. The amygdala function apparently knows nothing of place and time and is also a center for instinctive memory. The hippocampal function, on the other hand, provides spatial and temporal context for events. As long as there is cooperation between these two functions a person experiences a traumatic event as past and can remember its emotions in a spatial context as well. It was then at that place. But if there is only amygdalic impression without hippocampal qualification, the instinctual memory in that dissociation will lack context and the traumatic stress could come to presence at any time or place. The situation will be as though the traumatic event were not past whenever something triggers this timeless, placeless memory—the thunder might trigger the emotions I had when the bombs were exploding; your loving caress might unleash the terror of being violently hurt. When the amygdala's memory dominates, a person does not have a clear sense of having survived. It's not that one thing reminds a person of another thing. It's that without hippocampal function, the reminder lacks temporal and spatial identity.[2]

The higher the degree of hippocampal impairment due to traumatic stress, the lower the degree of control and management a person can exercise in relation to a traumatic experience. And the higher the degree of stress-induced impairment in the left cortical structure, which is largely responsible for speech—the more the terror is speechless—the lower the degree of ability for handling the stress constructively and meaningfully.

Just when we would like to make a productive difference in appropriations of traumatic events we find a troublesome fate: when brains and systems of nerves are damaged by too much stress, individuals have difficulty not only managing the effects of trauma but also making sense of them. In this case the body's faceless functioning comes to the fore; good sense and meaning fade away, and a physical dimension without intelligent, spatial, or temporal intention provides the traumatic presentation of a life. In such situations we are at the mercy of an interconnection of hormonal reactions and may well be closer to reptilian conditioning than to human sensibility. A trauma can be affectively recalled, for example, by a simple increase of heart rate or heavy breathing or the body's posture that is "reminiscent" of a traumatic occurrence.[3] And external triggers for traumatic stress can happen in benign situations by virtue of a color, sight, taste, touch, or smell. These instances of physical memory can be quite independent of what a person knows or wills.

Trauma happens as a physical presentation of danger that is sensed as life-threatening. The somatic aspect of traumatic occurrences is not much different in mice, alligators, and humans. In that presentation the ancient limbic system gains a definitive force. A body responds instinctively in the affects of defense, resistance, and mere urge to be. This dimension of physical presentation carries with it an affective memory that differentiates some sensations by their concurrence with the extreme danger. That differentiation is affected without temporal, spatial, or linguistic context. Limbic awareness is indifferent to such contextualization and specification. It—this dimension of awareness, this presentative force—differentiates sensed and extreme danger, tacitly remembers what its sensations are, and, when the memory traces are activated, triggers the hormonal, neurological chain that constitutes a major part of traumatic experiences. When a traumatized limbic system dominates, we have a degree of stress that overrides other affects, and we have a measure of sensation that is without the affections of reasonable or communal expression. It is affection with no sense of identity. In fact, human trauma seems to have in common a kind of wounding that ranges from shattering a sense of self to putting a sense of self in question. The shattering extremity appears to happen in part because of the uncontrolled magnitude of the event and in part because a sense of self plays hardly any role in the trauma's physical presentation.

Some of those who have undergone intense trauma speak of watching it happen to them as though they were outside of it—safely distanced—and as though they were articulating a vast indifference to themselves in the traumatic occurrence. That occurrence is there—I see it (strangely, it's happening to my body but not to me-seeing-it-happen. Who is that man drowning? Looks like me. I believe he's stopped breathing.) But that occurrence is not I. I'm elsewhere.

This distance of I from the trauma's immediacy has its survival value, I assume, in many situations. Or, if not survival, its value in a cerebral release from trauma's power. It allows a distance, in the drowning example, from the feel of the water entering my lungs, the terror of the heart's fluttering, pounding, and slowing, the effects of strangulation. Indeed, the blue of the water and the filtered rays of light, the increasingly slow motion of the body, the stilling of the water where there had been so much thrashing, the white sand rising up to the sinking thing, the light streaming down to darkness—there is something beautiful in the indifference of drowning. But now imagine a dark figure plunging in the water. Something jarring happens. Like a rude awakening. The distance collapses into terrible chest pain, heaving efforts to cough. The affections of vomiting, sucking air, water stinging and congesting air passages, flowing out. Unbearable pressure in my head. Agonizing light. I, having drowned, am now here.

Differentiation in Response to Traumatic Experiences

I want to note two ways that a dimension of indifference in trauma can continue after the initial event. One is found as people undergo affective disorders in which the stress of the trauma returns in ways that challenge or shatter the self's specific sense of itself. The limbic indifference to place and time that I've noted operates with its primitive memory in stress disorders, often with an extremity of force that makes normal daily living impossible. In these situations indifference appears, not as a distance in presentation but in blind inappropriateness for given circumstances and in a destructive noncoordination with the abilities of social consciousness and self-direction. Something without value one way or another appears—something without character or personality, without clarity of interest, intelligence, or choice. I am noting the mere indifference of an instinct severed from the partiality and interests of complex human awareness and values.

A second way in which a dimension of trauma's indifference appears is found by means of resilience and forgetting. Freud said that repression happens when a person refocuses attention away from unpleasant experiences, images, or thoughts. But whether such refocusing is a bad thing is open to question. When a person is able to focus attention on something futural and not dominated by a traumatic experience—when, that is, a person curbs the traumatic disturbance and, not so much excludes it actively from consciousness as simply moves on to an orientation that feels constructive—when a person is able to diminish the traumatized limbic force and define herself by other kinds of awareness, I expect that a good thing happens. This ability to forget might well assume a process of effective therapy. It might assume a strong sense of identity or a predisposition to let past events go. Whatever the conditions of renewed resilience that allows us to make good our losses, we can see that traumatic memory doesn't have to make a major difference in our lives.[4] It is there without differentiation, neutral in its disposition, available for a wide variety of differentiating appropriations. It is heedless in its limbic presence.

Traumatic memory might well make differences as it is appropriated in a nontraumatic context (I know that I can drown in a way I did not know before, for example; I might or might not want to swim again).[5] Or we might live without immediate reference to the traumatic experience. I speculate that limbic memory is variable in its persistence, that there are instances in which it is highly persistent and others where it fades or ceases. In any case, an indifferent dimension of trauma is awkwardly apparent as a life goes on without it. It is neutral as to value and to any form it might or might not take. It can be forgettable and without consequence in processes of living. Its memory is sometimes expendable. It need not be, and a person does not need to be with it. From a different angle we can say that the speechless, largely shapeless, careless, and impersonal dimension of traumatic experiences is, in its ineffability, rather more reptilian than divine and distinctly somatic in its uncanniness.

Traumatic Events and Their Wake

If a traumatic experience is definitively dependent on a speechless and utterly nonreflective physical dimension in our lives, what are we to say of the events that break traumatically into our lives? We have come so far as to see that the traumatized, limbic occurrence is not so much a narrative as a moment—often a recurring moment—of blind reaction, and this occurrence plays a major role in presenting traumatic events. The presentation of traumatic events is not at first one of narration. No one is telling a thing. There is simply a normal, blind and dumb reaction to a threat to a person's life. There are of course many stories to be told about the event and its presentation. Let's focus for now on the event and its presentation and see if we can continue to develop a narrative about them that brings out a dimension of them that is considerably different from the narrative aspect with its structure and meaning. Let's see if we can continue our narrative by speaking of instances of indifference to narrativity. I would like for a degree of resilience to happen in the process of this continuation—a counterpoint to traumatic stress—in which a breach in narrativity is presented without obsessional attachment to a past and without a sense of trauma attached to the breach in narrativity. This would be a breach that is attuned to the indifference we are addressing and a kind of resilience that happens in the affect of Mark Twain's words, "the secret source of humor itself is not joy but sorrow."

The untimely death of someone I love is not funny. Nor are the deaths of millions of people by murderous regimes. Nor are torture and many other suffering and deathly things, including drowning and abuse of infants. An event that threatens a life happens without humor or tragedy. It might threaten a person's life or an organism's life that is not yet a person, such as an infant. A traumatizing event might threaten the living fabric of a community or a system of belief. Such destructive threats can all constitute traumatic events. In any case, a violently threatening event simply happens, and when it happens it hits or threatens to hit something vital, something that urges to be. We may call that a limbic hit. Whether or not the event is recognized with reflective awareness, something nonreflective is activated. At this level the threatening event's sensed violence incites a nonverbal intervention in people's worlds of meaning. A body perceives a threat of violence, perhaps by inference, perhaps by a completely unexpected impact. The event that incites, incites by impact, by intrusion, by destabilizing expectations and harmonies, by dissociational disturbance. Part of its wounding force is in the event's refusal of operating rules, its shock value. Of course, an event that is traumatic for me might fit well and with no hint of trauma in another person's or culture's life—trauma is not wed necessarily to any particular context. But when it hits, it hits outside of captivity by whatever is expected as usual and right—and outside of the procedures of control that are necessary for a narrative. The indifference of pastness and distance is needed for that kind of control.

A sympathy between such events and limbic speechlessness seems to take place. Neither is a story or like a story. Each in its immediacy requires language to remain at a distance. Traumatic events and limbic reaction, in their violence and senselessness, allow that distance, that incomprehensibility, that presence without reason that puts human subjects on edge and sends some of us spinning in obsessional efforts to regain or reflect an impossible proximity to hidden sense. Others react by efforts to invest in traumatic events a patina of hidden divinity or a hope for possible benefit. An event of trauma is as indifferent to human values as the agitated limbic system is. We can turn it virtually by stories and concepts to our values and benefits. But in conceiving a traumatic event we are conceiving something already there and gone, conceptless, violent, speechless, senseless, dangerous. The traumatic event is in its life simply its violent happening, reducible to nothing.

So fine. That's trauma all right. That's its event. Often there is so much that is lost—parts of bodies, whole bodies, happiness, loved people, locales, societies, innocence, hope, affection, a sense of self. Everything sooner or later. Probably no one escapes being traumatized—it's a tristful story, full of sorrow and punctuated by losses that no words or intentions can capture. And yet, in wanting to be alive, we, in effect, ask for it. We intend to make good our losses. We begin again. We might exaggerate our happiness and confidence and creativity. Sometimes we whistle in the dark and smile when we reach a well-lighted place. We try. Sometimes we exaggerate our losses or cultivate them. We build new systems and write new rules. Unless we are beset by discouragement, and then, if not overwhelmed, we talk. We talk with our friends, our therapists, ourselves. Maybe we look for sympathy.

Often, for understanding. We cry and grieve. We drag ourselves through days and nights. We build a new sense of who we are now. We look toward a time when we don't want to quit and will want to find something worthwhile to do. Mostly we want to live well.

I think this aim to live well is part of what Twain had in mind when he said that humor arises from sorrow. I think he knew that he could dwell on his losses, write of them, allow them to control his stories and affections, give the awful speechless a certain mastery of his speech. And he knew, I believe, as he moved on in his life, that there is a diminished future in projects that continually return to past losses. He knew sorrow too well. So, he made a different turn. He often said, in effect, "Did I tell you about the billy-goat by the name of Eugene who could never turn down the opportunity presented when someone bent over to pick something up? People said he kept count of how far he had knocked bodies flying and that his aim was to put the heaviest thing around the farm flat over the barnyard fence. Well, one day the farmer's wife, Mrs. Bucklebee, took her basket off the peg on the back porch and went to the barnyard to find the eggs of a setting hen. It was a day that Eugene felt like a challenge to cure his irritability and help his outlook on life. He looked around the corner of the shed where he hung out when it was hot and spied Mrs. Bucklebee coming along humming to herself. He closed one eye, wrinkled the other, and commenced to figure her weight. She . . ." and on he would go into his story.

Perhaps that is repression. I think it's probably better called making good his losses, a kind of forgetting that Nietzsche recognized as life-enhancing. It's presenting traumas and sorrow—like the ones Twain lived through—with antiphrasis, with humor and in an absence of big hope—with the intention, as he has Adam say of Sabbaths with Eve in the Garden of Eden, of making it through to Monday.

Notes

This chapter first appeared in Charles Scott, *Living with Indifference* (Bloomington, IN: Indiana University Press, 2007), 125–34.

1. For these remarks on the physiology of trauma I will draw from Babette Rothschild's helpful report and discussion in *The Body Remembers* (New York: Norton, 2000), esp. pt. 1.

2. This atemporal and aspatial memory has a particular importance for understanding the effects of certain kinds of infantile trauma. Rothschild summarizes nicely: "[T]he amygdala is mature at birth and . . . the hippocampus matures later, between the second and third year of life. Understanding the difference in the maturational schedules, as well as the functions of these two structures, provides an explanation for the phenomenon of infantile amnesia—

the fact that we usually don't consciously remember our infancy. Infantile experiences are processed through the amygdala on the way to storage in the cortex. Hippocampal function is not yet available, so the resulting memory of an infantile experience includes emotion and physical sensations without context or sequence. This is the probable explanation for why, in later life, infantile experiences cannot be accessed as what we usually call memories" (ibid., 21).

3. Ibid., 36.

4. I take the phrase "make good our losses" from "In the Wake of Darwin" by Vincent Colapietro, in *In Dewey's Wake: Unfinished Work of Pragmatic Reconstruction*, ed. W. J. Gavin (New York: State University of New York Press, 2003), 215. He also includes this quotation from Adam Phillips, "Life was about what could be done with what was left, with what still happened to be there" (232).

5. This drowning example is not entirely based on my own experience. Although I have had two slightly traumatic experiences in water in which the limbic system and autonomic nervous system took over, I did not drown. I have spoken with one person who did drown and was brought back to life, and I have a secondhand report from a distant relative who spoke of his experience of drowning and recovery.

Part 3

Trauma and Clinical Approaches

7

Crime and Memory

Judith Lewis Herman, M.D.

The conflict between knowing and not knowing, speech and silence, remembering and forgetting, is the central dialectic of psychological trauma. This conflict is manifest in the individual disturbances of memory, the amnesias and hypermnesias, of traumatized people. It is manifest also on a social level, in persisting debates over the historical reality of atrocities that have been documented beyond any reasonable doubt. Social controversy becomes particularly acute at moments in history when perpetrators face the prospect of being publicly exposed or held legally accountable for crimes long hidden or condoned. This situation obtains in many countries emerging from dictatorship, with respect to political crimes such as murder and torture. It obtains in this country with regard to the private crimes of sexual and domestic violence. This chapter examines a current public controversy, regarding the credibility of adult recall of childhood abuse, as a classic example of the dialectic of trauma.

What happens to the memory of a crime? What happens to the memory in the mind of the victim, in the mind of the perpetrator, and in the mind of the bystander? When people have committed or suffered or witnessed atrocities, how do they manage to go on living with others, in a family, in a community, and how do others manage to go on living closely with them?

This is the question I propose to explore. As my starting point, I would like to recount a case reported by Dan Bar-On, an Israeli psychologist who has investigated the generational impact of the Nazi Holocaust. Bar-On has done extensive interviews not only with children of Holocaust survivors, but also with children of the Nazi SS. In fact, for some years now, he has been conducting workshops, in both Israel and Germany, in which he brings members of these two groups together. In these workshops, the children of victims and the children of perpetrators disclose to one another the stories of the crimes that their families kept secret. Such encounters represent the highest form of therapeutic endeavor, for they carry the potential for both personal and social healing.

During the mid-1980s, Bar-On interviewed forty-eight men and women whose fathers (and, in one case, a mother) had participated either directly or indirectly in extermination activities during World War II. He asked them to recall whether their parents had ever discussed wartime experiences at home and whether they had shown any signs of guilt, regret, or moral conflict. Recognizing that to address such questions would be emotionally stressful for both his subjects and himself, he took care to build rapport and trust with his subjects, and to maintain his own institutional, collegial, and personal support. No one can do this kind of work alone.

The adult children of Nazi war criminals could not initially remember any discussion whatsoever in their families, either of the extermination program in general or of their parents' participation. They also reported that they saw little evidence of distress or moral conflict in their parents. They dealt with the problem of their lack of knowledge by repeatedly using one sentence that may sound all too familiar: "We had a very normal family life."

Some of the adult children constructed their own version of historical events from small bits of information they had gleaned from various sources, minimizing the role their fathers had played. One man explained that his father had been a train driver during the war, but only drove ammunition transports, and had never personally transported Jews to the death camps. When Bar-On expressed skepticism, on the basis of well-established historical evidence, this man agreed to ask his father for more information. For the first time in his life, he asked his father direct questions about the past; a few days later he recounted their conversation to Bar-On. At first he reiterated the original story: his father denied any involvement in the transport of Jews and had not known anything about it. On further inquiry, he said that his father had admitted hearing about it from others at the time. Just as the interview was about to end, he suddenly added: "And this time, my father told me of another matter. He was on duty when they took a big group of prisoners of war and shot them on the platform in front of his eyes."

"How terrible!" Bar-On exclaimed. "It must have been very difficult to keep that hidden all these years."

"This was the first time he spoke to me about it," the son replied, matter-of-factly. "He never told anyone about it."

A year later, Bar-On reinterviewed the same informant. The memory that had been recovered in the previous interview was gone. The man did not remember his father's disclosure, or that he had in turn repeated the story to Bar-On. Reflecting on this case, Bar-On invoked the image of a double wall erected to prevent acknowledgment of the memory of crime. The fathers did not want to tell; the children did not want to know.[1]

The ordinary human response to atrocities is to banish them from consciousness. Certain violations of the social compact are too terrible to utter aloud; this is the meaning of the word *unspeakable*. Atrocities, however, refuse to be buried. As powerful as the desire to deny atrocities is the conviction that denial does not work. Our folk wisdom and classic literature are filled with ghosts who refuse to rest in their graves until their stories are told, ghosts who appear in dreams or visions, bidding their children, "Remember me." Remembering and telling the truth about terrible events are essential tasks for both the healing of individual victims, perpetrators, and families and for the restoration of the social order.

The conflict between the will to deny horrible events and the will to proclaim them aloud is the central dialectic of psychological trauma. I would like to explore the impact of this dialectic on the phenomenon of remembering. I will first address what perpetrators remember, and then what victims remember, and finally—and this is perhaps the most complicated task of all—what bystanders and witnesses remember.

What do perpetrators remember? Here our professional ignorance is almost perfect. We know so very little about the inner lives of people who commit atrocities, that relatively sophisticated investigations, such as studies of memory, are utterly beyond our current capability. We know so little about perpetrators, first, because they have no desire for the truth to be known; on the contrary, all observers agree on their deep commitment to secrecy and deception. Perpetrators are not generally friendly to the process of scientific inquiry. Usually they are willing to be studied only when they are caught, and under those circumstances they tell us whatever it is they think we want to know. In general, we have wanted to know very little. The dynamics of human sadism have almost entirely escaped our professional attention. Our diagnostic categories do not comprehend the perpetrators; they present an appearance of normality, not only to their children, but also to us.

By contrast, we now know a fair amount about what victims remember. It seems clear that close-up exposure, especially early and prolonged exposure, to human cruelty has a profound effect on memory. Disturbances of memory are a cardinal symptom of posttraumatic disorders. They are found equally in the casualties of war and political oppression: combat veterans, political prisoners, and concentration camp survivors; and in the casualties of sexual and domestic oppression: rape victims, battered women, and abused children. These disturbances have been difficult to comprehend because they are apparently contradictory. On the one hand, traumatized people remember too much; on the other hand, they remember too little. They seem to have lost "authority over their memories" (I borrow the phrase from my colleague Mary Harvey).[2] The memories intrude when they are not wanted, in the form of nightmares, flashbacks, and behavioral reenactments. Yet the memories may not be accessible when they are wanted. Major parts of the story may be missing, and sometimes an entire event or series of events may be lost. We have by now a very large body of data indicating that trauma simultaneously enhances and impairs memory. How can we account for this? If traumatic events are (in the words of Robert J. Lifton) "indelibly imprinted,"[3] then how can they also be inaccessible to ordinary memory?

When scientific observations present a paradox, one way of resolving the contradiction is to ignore selectively some of the data. Hence, we find some authorities even today asserting that traumatic amnesia cannot possibly exist because, after all, traumatic events are strongly remembered. Fortunately for the enterprise of science, empirical observations do not go away simply because simplistic theories fail to explain them. On the contrary, I believe that some of our most important discoveries arise from attempts to understand apparent paradoxes of this kind. I would like to offer two theoretical constructs that may help us clarify and organize our thinking in this area. The first is the concept of state-dependent learning; the second is the distinction between storage and retrieval of memory.

The common denominator—the A criterion—of psychological trauma is the experience of terror. Traumatic events are those that produce "intense fear, helplessness, loss of control, and threat of annihilation."[4] This is the definition in the fourth edition of the *Comprehensive Textbook of Psychiatry*, and extensive studies in the DSM-IV field trials have essentially confirmed this observation. People in a state of terror are not in a normal state of consciousness. They experience extreme alterations in arousal, attention, and perception. All of these alterations potentially affect the storage and retrieval of memory.

The impact of hyperarousal on memory storage can be studied in the laboratory with animal models. James McGaugh and his colleagues have demonstrated in an elegant series of experiments that high levels of circulating catecholamines result in enhanced learning that stubbornly resists subsequent extinction.[5] This is an animal analogue, if you will, of the "indelible imprint" of traumatic events on memory. Building on McGaugh's concept of overconsolidated memory, Roger Pitman and his colleagues have demonstrated that activation of trauma-specific memories in combat veterans with post-traumatic stress disorder (PTSD) produces highly elevated physiologic responses that fail to extinguish even over periods of half a lifetime.[6] They interpret their findings as evidence for overconsolidation of memories laid down in a biological state of hyperarousal.

When people are in a state of terror, attention is narrowed and perceptions are altered. Peripheral detail, context, and time sense fall away, while attention is strongly focused on central detail in the immediate present.

When the focus of attention is extremely narrow, people may experience profound perceptual distortions including insensitivity to pain, depersonalization, derealization, time slowing, and amnesia. This is the state we call dissociation. Similar states can be induced voluntarily through hypnotic induction techniques, or pharmacologically, with ketamine, a glutamate receptor antagonist.[7] Normal people vary in their capacity to enter these altered states of consciousness.

Traumatic events have great power to elicit dissociative reactions. Some people dissociate spontaneously in response to terror. Others may learn to induce this state voluntarily, especially if they are exposed to traumatic events over and over. Political prisoners instruct one another in simple self-hypnosis techniques in order to withstand torture. In my clinical work with incest survivors, again and again I have heard how, as children, they taught themselves to enter a trance state.

These profound alterations of consciousness at the time of the trauma may explain some of the abnormal features of the memories that are laid down. It may well be that because of the narrow focusing of attention, highly specific somatic and sensory information may be deeply engraved in memory, whereas contextual information, time-sequencing, and verbal narrative may be poorly registered. In other words, people may fail to establish the associative linkages that are part of ordinary memory.

If this were so, we would expect to find abnormalities not only in storage of traumatic memories, but also in retrieval. On the one hand, we would expect that the normal process of strategic search, that is, scanning autobiographical memory to create a coherent sequential narrative, might be relatively ineffective as a means of gaining access to traumatic memory. On the other hand, we would expect that certain trauma-specific sensory cues, or biologic alterations that reproduce a state of hyperarousal, might be highly effective. We would also expect that traumatic memories might be unusually accessible in a trance state.

This is, of course, just what clinicians have observed for the past century. The role of altered states of consciousness in the pathogenesis of traumatic memory was discovered independently by Janet and by Breuer and Freud one hundred years ago. The concepts of state-dependent memory and abnormal retrieval were already familiar to these great investigators. Indeed, it was Janet who first coined the term "dissociation."[8] More recently, civilian disaster studies, notably those by David Spiegel and his colleagues,[9] have demonstrated that people who spontaneously dissociate at the time of the traumatic event are the most vulnerable to developing symptoms of PTSD, including the characteristic disturbances of memory retrieval: intrusive recall and amnesia.

Abnormal memory retrieval in posttraumatic disorders has also now been demonstrated in the laboratory. This is a very fertile and exciting area of current investigation. For example, a research team at Yale University has been able to induce flashbacks in combat veterans with PTSD using a yohimbine challenge; the same effect could not be produced in veterans who did not have PTSD.[10] Studies of traumatized people now demonstrate that some have abnormalities not only in trauma-specific memory but also in general memory. Richard McNally and his colleagues have noted that combat veterans with PTSD have difficulty retrieving specific autobiographical memories, especially after being exposed to a combat videotape. As they interviewed their subjects in the laboratory, McNally and his colleagues were struck by the fact that the men who showed the greatest disturbances in autobiographical memory were those who still dressed in combat regalia twenty years after the war.[11] These men remembered nothing in words and everything in action. The contemporary researchers had rediscovered what was already well-known to the great nineteenth-century clinical investigators, namely, that traumatic memories could manifest in disguised form as somatic and behavioral symptoms. Janet attributed the symptoms of hysteria to "unconscious fixed ideas."[12] Breuer and Freud wrote that "hysterics suffer mainly from reminiscences."[13]

This puzzling and fascinating phenomenon has been extensively documented in contemporary clinical studies as well. For example, among twenty children with documented histories of early trauma, Lenore Terr found that none could give a verbal description of the events that had occurred before they were two and one-half years old. Nonetheless, these experiences were indelibly encoded in memory and expressed nonverbally, as symptoms. Eighteen of the twenty children showed evidence of traumatic memory in their behavior and their play. They had specific fears and somatic symptoms related to the traumatic events, and they reenacted these events in their play with extraordinary accuracy. A child who had been sexually molested by a babysitter in the first two years of life could not, at age five, remember or name the babysitter. Furthermore, he denied any knowledge or memory of being abused. But in his play he repeatedly enacted scenes that exactly replicated a pornographic movie made by the babysitter. This highly visual and enactive form of memory, appropriate to young children, seems to be mobilized in adults as well as in circumstances of overwhelming terror.[14]

In Bessel van der Kolk's phrase, "the body keeps the score."[15] Traumatic memories persist in disguised form as psychiatric symptoms. The severity of symptoms is highly correlated with the degree of memory disturbance. Data from numerous clinical studies including DSM-IV field trials for PTSD now demonstrate a very strong correlation between somatization, dissociation, self-mutilation, and other self-destructive behaviors, and childhood histories of prolonged, repeated trauma.[16]

Although it is clear by now that abnormalities of memory are characteristic of posttraumatic disorders, they are not seen in all traumatized people, even after the most catastrophic exposure. For example, in a community study of refugee survivors of the Cambodian genocide, Eve Carlson found that 90 percent reported some degree of amnesia for their experiences but 10 percent did not.[17] In childhood abuse survivors, we now have several clinical studies and two community studies. Memory disturbances seem to fall on a continuum, with some subjects reporting that they always remembered the traumatic events, some reporting partial amnesia with gradual retrieval and assimilation of new memories, and some reporting a period of global amnesia, often followed by a period of intrusive and highly distressing delayed recall. The percentage of subjects falling into this last category ranges from 26 percent in a study I conducted with my colleague Emily Schatzow,[18] to 19 percent in a more recent study by Loftus, Polonsky, and Fullilove.[19] The degree of amnesia may be correlated with the age of onset, duration, and degree of violence of the abuse. Further research is needed to clarify both the determinants of the memory disturbance and the mechanism of delayed recall.

The nineteenth-century investigators not only documented the role of traumatic memory in the pathogenesis of hysterical symptoms, but also found that these symptoms resolved when the memories, with their accompanying intense affect, were reintegrated into the ongoing narrative of the patient's life. These discoveries are the foundation of modern psychotherapy. "Memory," Janet wrote, "like all psychological phenomena, is an action; essentially it is the action of telling a story. . . . A situation has not been satisfactorily liquidated . . . until we have achieved, not merely an outward reaction through our movements, but also an inward reaction though the words we address to ourselves, through the organization of the recital of the event to others and to ourselves, and through the putting of this recital in its place as one of the chapters in our personal history."[20]

Throughout the next century, with each major war, psychiatrists who treated men in combat rediscovered this same therapeutic principle. They found that traumatic memories could be transformed from sensations and images into words, and that when this happened, the memories seemed to lose their toxicity. The military psychiatrists also rediscovered the power of altered states of consciousness as a therapeutic tool for gaining access to traumatic memories. Herbert Spiegel pioneered the use of hypnosis with acutely traumatized soldiers in World War II.[21] Roy Grinker and John Spiegel used sodium amytal.[22] These psychiatrists understood, however, that simple retrieval of memory was not sufficient in itself for successful treatment. The purpose of therapy was not simply catharsis, but rather integration of memory.

Those of us who treat civilian casualties of sexual and domestic violence have had to rediscover these same principles of treatment. Retrieval of traumatic memory, in the safety of a caring relationship, can be an important component of recovery, but it is only one small part of the "action of telling a story." In this slow and laborious process, a fragmented set of wordless, static images is gradually transformed into a narrative with motion, feeling, and meaning. The therapist's role is not to act as a detective, jury, or judge, not to extract confessions or impose interpretations on the patient's experience, but rather to bear witness as the patient discovers her own truth. This is both our duty and our privilege.

In my review of the current state of the field, it may be noticed that I have not said anything about the accuracy or verifiability of traumatic memories. It has been widely presumed that traumatic memories, especially those retrieved after a period of amnesia, might be particularly prone to distortion, error, or suggestion. In fact, a careful review of the relevant literature yields the conclusion that traumatic memories may be either more or less accurate than ordinary memories, depending on which variables are studied. For example, such memories may be generally accurate, or better than accurate, for gist and for central detail. They may be quite inaccurate when it comes to peripheral detail, contextual information, or time sequencing.[23]

On the matter of verifiability, there are some fascinating single case reports of traumatic memories from childhood, retrieved after a period of dense amnesia and later confirmed beyond a reasonable doubt.[24] These anecdotal reports prove only that such memories can turn out to be true and accurate; they do not permit us to draw any conclusions about how reliable such memories might be in general. I know of only two systematic studies in which subjects were asked whether they knew of evidence to confirm their memories of childhood trauma. The first is the clinical study Emily Schatzow and I conducted with fifty-three incest survivors in group therapy.[25] The majority of these patients undertook an active search for information about their childhood while they were in treatment. As a result, 74 percent were able to obtain some form of verification. More recently, Feldman-Summers and Pope conducted a nationwide study of 330 psychologists.[26] Of these, 23.9 percent gave a history of childhood physical or sexual abuse, a figure consistent with general community surveys. Exactly half of these subjects reported that they had some independent source of information corroborating their memories. In these two studies, the subjects who reported amnesia and delayed recall did not differ from those with continuous memory in their ability to obtain confirming evidence. The limitations of these studies should be noted, however; because these were not forensic investigations, the researchers did not independently confirm the subjects' reports.

Finally, I know of no empirical studies indicating that people who report histories of trauma are any more suggestible, or more prone to lie, fantasize, or confabulate, than the general population. Nevertheless, whenever survivors come forward, these questions are inevitably raised. In the absence of any systematic data, those who challenge the credibility of survivors' testimony repeatedly resort to argument from anecdote, overgeneralization, selective omission of relevant evidence, and frank appeals to prejudice. The cry of "witch hunt" is raised, invoking an image of packs of irrational women bent on destroying innocent people. When this happens, we must recognize that we have left the realm of scientific inquiry and entered the realm of political controversy.[27]

This brings us to the final subject: When a crime has been committed, what do the bystanders remember? For we are the bystanders, and we are called on to bear witness to the many crimes that occur, not far away in another time and place, but in our own society, in normal families very much like our own, perhaps in our own families. Like the son of the man who drove the trains in wartime, we have been reluctant to know about the crimes we live with every day. We have sought information only when prodded to do so, and once we have acquired the information we have been eager to forget it again as soon as possible. We can see the phenomenon of active forgetting in operation as it pertains to crimes against humanity carried out on the most massive scale of organized genocide. It operates with the same force in the case of those unwitnessed crimes carried out in the privacy of families.

When we bear witness to what victims remember, we are inevitably drawn into the conflict between victim and perpetrator. Although we strive for therapeutic neutrality, it is impossible to maintain moral neutrality. To clarify the difference: therapeutic neutrality means remaining impartial with regard to the patient's inner conflicts, respecting his or her capacity for insight, autonomy, and choice. This is a cardinal principle of all psychotherapy and is of particular importance in the treatment of traumatized people, who are already suffering as the result of another's abuse of power. Moral neutrality, by contrast, means remaining impartial in a social conflict. When a crime has been committed, moral neutrality is neither desirable nor even possible. We are obliged to take sides. The victim asks a great deal of us; if we take the victim's side, we will inevitably share the burden of pain and responsibility. The victim demands risk, action, engagement, and remembering. The perpetrator asks only that we do nothing, thereby appealing to the universal desire to see, hear, and speak no evil, the desire to forget.

In order to escape accountability for their crimes, perpetrators will do everything in their power to promote forgetting. Secrecy and silence are the perpetrator's first lines of defense, but if secrecy fails, the perpetrator will aggressively attack the credibility of the victim and anyone who supports the victim. If the victim cannot be silenced absolutely, the perpetrator will try to make sure that no one listens or offers aid. To this end, an impressive array of arguments will be marshaled, from the most blatant denial to the most sophisticated rationalizations. After every atrocity one can expect to hear the same apologies: it never happened; the victim is deluded; the victim lies; the victim fantasizes; the victim is manipulative; the victim is manipulated; the victim brought it on himself or herself (masochistic); the victim exaggerates (histrionic), and, in any case, it is time to forget the past and move on. The more powerful the perpetrator, the greater will be his prerogative to name and define reality, and the more completely his arguments will prevail.

This is what has happened in our profession. In the past we have been only too ready to lend our professional authority to the perpetrator's version of reality. For decades we taught that sexual and domestic crimes are rare, when in fact they are common; for decades we taught that false complaints are common, when in fact they are rare. At times, we have been willing to see what happens to men assaulted on the battlefield and women and children assaulted in the home. But we have been unable to sustain our attention for very long. The study of psychological trauma has had a discontinuous history of our profession. Periods of active investigation have alternated with periods of oblivion, so that the same discoveries have had to be made over and over again.

Why this curious amnesia? The subject of psychological trauma does not languish for lack of scientific interest. Rather, it provokes such intense controversy that it periodically becomes anathema. Throughout the history of the field, dispute has raged over whether patients with posttraumatic conditions are entitled to care and respect or deserve contempt, whether they are genuinely suffering or malingering, whether their histories are true or false, and, if false, whether imagined or maliciously fabricated. Despite a vast body of literature empirically documenting the phenomena of psychological trauma, debate still centers on the most basic question: whether these phenomena are credible and real.

It is not only the patients but also the investigators of posttraumatic conditions whose credibility has been repeatedly challenged. Clinicians and researchers who have listened too long and too carefully to traumatized patients have often become suspect among their colleagues, as though contaminated by contact. Investigators in this field have often been subjected to professional isolation. Most of us are not very brave. Most of us would rather live in peace. When the price of attending to victims gets to be too high (and recently, we learned that the price can be as high as half a million dollars),[28] most of us find good reasons to stop looking, stop listening, and start forgetting.

We find ourselves now at a historic moment of intense social conflict over how to address the problem of sexual and domestic violence. In the past twenty years, the women's movement has transformed public awareness of this issue. We are now beginning to understand that the subordination of women is maintained not only by law and custom, but also by force. We are beginning to understand that rape, battery, and incest are human rights violations; they are political crimes in the same sense that lynching is a political crime, that is, they serve to perpetuate an unjust social order through terror.[29] The testimony of women, first in the privacy of small groups, then in public speak-outs, and finally in formal epidemiologic research, has documented the fact that these crimes are common, endemic, and socially condoned. Grassroots activists pioneered new forms of care for victims (the rape crisis center and the battered women's shelter), and advocated for legal reforms that would permit victims to seek justice in court. As a result, we now find ourselves in a situation where for the first time perpetrators face the prospect of being held publicly accountable.

I should emphasize the fact that the odds still look very good for perpetrators. Most victims still either keep the crime entirely secret or disclose only to their closest confidantes. Very few take the risk of making their complaints public. The most recent data we have indicate that although the reporting rate for rape may have doubled in the last decade, it is still only 16 percent.[30] For sexual assaults on children, the rate is even lower, ranging from 2 to 6 percent.[31] These numbers are further reduced at each step along the way to trial. Victims of sexual and domestic crimes still face an uphill battle in court. Besides the strong constitutional protections that all defendants enjoy (and which no one is proposing to abrogate), perpetrators are also aided by the widespread bias against women that still pervades our system of justice. Nevertheless, even the prospect of accountability is extremely threatening to those who have been accustomed to complete impunity.

When people who have abused power face accountability, they tend to become very aggressive. We can see this in the political experience of countries emerging from dictatorships in Latin America or in the former Soviet bloc. In many cases the military groups or political parties that were responsible for human rights violations retain a great deal of power, and they will not tolerate any settling of accounts. They threaten to retaliate fiercely against any form of public testimony. They demand amnesty, a political form of amnesia.[32] Faced with exposure, the dictator, the torturer, the batterer, the rapist, the incestuous father all issue the same threat: if you accuse me I will destroy you and anyone who harbors or assists you.

This social conflict over accountability has reached a peak of intensity just at the same moment that we in the mental health professions are struggling to relearn and integrate the fundamental principles of diagnosis and treatment of traumatic disorders. We professionals are just now feeling the backlash that grassroots workers in women's and children's services have already endured for quite some time. Just as mental health professionals are starting to figure out how to treat survivors (often by trial and error), we suddenly find ourselves and the work we do under very serious attack. Some of these attacks are funny; some are quite ugly. Most of us are not accustomed to threatening phone calls, pickets in front of our homes or offices, entrapment attempts, or legal harrassment; but we are going to have to learn fast how to cope with these and other intimidation tactics.

We have three choices. We can ally with, and become apologists for, accused perpetrators, as some distinguished authorities have done. We can back away from the whole field of traumatic disorders, as has happened many times in the past.[33] Or we can determine not to give in to fear, but rather to continue our work—in the laboratory, in the privacy of the consulting room, and ultimately in public testimony.

We need to be clear about the nature of the work that we do. The pursuit of truth in memory takes different forms in psychotherapy, where the purpose is to foster individual healing; in scientific research, where the purpose is to subject hypotheses to empirical test; and in court, where the purpose is to mete out justice. Each setting has a different set of rules and standards of evidence, and it is important not to confuse them. It is no more appropriate to apply courtroom procedures and standards of evidence in the consulting room or the laboratory than to apply therapeutic or laboratory procedures and standards of evidence in the courtroom. But if we pursue the truth of memory in scientific and therapeutic setting, then we will inevitably have to defend our work in the courtroom as well. For our work places us in the role of the bystander, bearing witness to the memory of crimes long hidden. Some of our patients will eventually choose to seek justice. Our stance regarding this decision should be one of technical neutrality. Nowhere is the principle of informed choice more important. When I am consulted I always suggest that patients think long and hard about the consequences of taking this step; it is not a decision to be made impulsively. But when, after careful reflection, some of our patients choose to speak publicly and to seek justice, we will be called on to stand with them. I hope we can show as much courage as our patients do. I hope that we will accept the honor of bearing witness and stand with them when they declare: we remember the crimes committed against us. We remember, we are not alone, and we are not afraid to tell the truth.

Notes

Thanks to the *Bulletin of the Academy of Psychiatry and Law* for permission to reprint this chapter. The article first appeared in *Bulletin of the Academy of Psychiatry and Law* 23, no. 1 (1995): 5–17 and was presented as the Manfred F. Guttmacher Award Lecture at the annual meeting of the American Psychiatric Association and the American Academy of Psychiatry and the Law on May 22, 1994, in Philadelphia, PA.

1. Daniel Bar-On, "Holocaust Perpetrators and Their Children: A Paradoxical Morality," *Journal for Humanistic Psychology* 29 (1989): 424–43.

2. Mary Harvey and Judith L. Herman, "Amnesia, Partial Amnesia and Delayed Recall among Adult Survivors of Childhood Trauma," *Consciousness and Cognition* 3 (1994): 295–306.

3. Robert J. Lifton, "The Concept of the Survivor," in *Survivors, Victims, and Perpetrators: Essays on the Nazi Holocaust*, ed. Joel E. Dimsdale (New York: Hemisphere, 1980), 113–26.

4. Nancy C. Andreasen, "Post-traumatic Stress Disorder," in *Comprehensive Textbook of Psychiatry*, ed. Harold I. Kaplan and B. J. Sadock, 4th ed. (Baltimore: Williams and Wilkins, 1985), 918–24.

5. James L. McGaugh, "Affect, Neuromodulatory Systems, and Memory Storage," in ed. Sven Ake Christianson, *The Handbook of Emotion and Memory* (Hillsdale, NJ: Erlbaum, 1992): 245–68.

6. Roger K. Pitman and Scott P. Orr, "Psychophysiology of Emotional Memory Networks in Post-Traumatic Stress Disorder," in *Proceedings of the Fifth Conference on the Neurobiology of Learning and Memory, University of California, Irvine, October 22–24, 1992* (London: Oxford University Press, 1995).

7. John H. Krystal, Laurence P. Karper, John P. Seibyl et al., "Subanesthetic Effects of the Noncompetitive NMDA Antagonist, Ketamine, in Humans: Psychotomimietic, Perceptual, Cognitive and Neuroendocrine Responses," *Archives of General Psychiatry* 51 (1994): 199–214.

8. Pierre Janet, *L'Automatisme psychologique: Essai de psychologie expérimentale sur les formes inférieures de l'activité humaine* (1889) (Paris: Payot, 1973).

9. Cheryl Koopman, Catherine Classen, and David Spiegel, "Loss of Home, Dissociation, and Stressful Life Change," paper presented at the 147th Annual Meeting of the American Psychiatric Association, May 23, 1994, Philadelphia, PA.

10. J. Douglas Bremner, Michael Davis, S. M. Southwick, J. H. Krystal, and D. S. Charney, "The Neurobiology of Post-traumatic Stress Disorder," *Review of Psychiatry* 12 (1993): 182–204.

11. Sharon B. Zeitlin and R. J. McNally, "Implicit and Explicit Memory Bias for Threat in Post-Traumatic Stress Disorder," *Behaviour Research and Therapy* 29 (1991): 451–57.

12. Pierre Janet, "Étude sur un cas d'aboulie et d'idées fixes," (1891) in *The Discovery of the Unconscious*, trans. Henri Ellenberger (New York: Basic Books, 1970), 365–66.

13. Joseph Breuer and Sigmund Freud, "Studies on Hysteria," (1893–95) in *The Standard Edition of the Complete Psychological Works of Sigmund Freud*, vol. 2 (London: Hogarth Press, 1962).

14. Lenore C. Terr, "What Happens to Early Memories of Trauma? A Study of Twenty Children under Age Five at the Time of Documented

Traumatic Events," *Journal of the American Academy of Child and Adolescent Psychiatry* 27 (1988): 96–104.

15. Bessel Van der Kolk, "The Body Keeps the Score: Memory and the Evolving Psychobiology of Post-Traumatic Stress Disorder," *Harvard Review of Psychiatry* 1 (1994): 253–65.

16. Bessel Van der Kolk, S. Roth, D. Pelcovitz, and S. Mandel, *Complex Post-Traumatic Stress Disorder: Results from the DSM-IV Field Trial for PTSD* (Washington, DC: American Psychiatric Association, 1993).

17. E. B. Carlson and R. Rosser-Hogan, "Cross-cultural Response to Trauma: A Study of Traumatic Experiences and Post-traumatic Symptoms in Cambodian Refugees," *Journal of Traumatic Stress* 7 (1994): 43–58.

18. Judith L. Herman and Emily Schatzow, "Recovery and Verification of Memories of Childhood Sexual Trauma," *Psychoanalytic Psychology* 4 (1987): 1–14.

19. Elizabeth F. Loftus, S. Polonsky, and M. T. Fullilove, "Memories of Childhood Sexual Abuse: Remembering and Repressing," *Psychology of Women Quarterly* 18 (1994): 67–84.

20. Pierre Janet, *Psychological Healing: A Historical and Clinical Study*, vol. 1, (1919), trans. Eden Paul and Cedar Paul (New York: Macmillan, 1925), 661.

21. Abram Kardiner and Herbert Spiegel, *The Traumatic Neuroses of War* (New York: Hoeber, 1941).

22. Roy R. Grinker and John P. Spiegel, *Men Under Stress* (Philadelphia: Blakiston, 1945).

23. See Sven Ake Christianson and Elizabeth Loftus, "Remembering Emotional Events: The Fate of Detailed Information," in *Cognition and Emotion* 5 (1991): 81–108; and A. Burke, F. Heuer, and D. Reisberg, "Remembering Emotional Events," *Memory and Cognition* 20 (1992): 277–90.

24. See, for example, Nathan M. Szajnberg, "Recovering a Repressed Memory, and Representational Shift in an Adolescent," *Journal of the American Psychoanalytical Association* 42 (1993): 711–27.

25. Herman and Schatzow, "Recovery."

26. Shirley Feldman-Summers and Kenneth S. Pope, "The Experience of 'Forgetting' Childhood Abuse: A National Survey of Psychologists," *Journal of Consulting and Clinical Psychology* 62, no. 3 (1994): 636–39.

27. Judith L. Herman, "Presuming to Know the Truth," *Nieman Reports* 48 (1994): 43–45.

28. B. D. Ayres, "Father Accused of Incest Wins Suit against Memory Therapists," *New York Times*, May 15, 1994, p. 29.

29. Susan Brownmiller, *Against Our Will: Men, Women, and Rape* (New York: Simon and Schuster, 1975).

30. Dean G. Kilpatrick and C. L. Best, *Rape in America: A Report to the Nation* (Arlington, VA: National Victim Center, 1992).

31. Diana E. H. Russell, *Sexual Exploitation: Rape, Child Sexual Abuse, and Workplace Harassment* (Beverly Hills, CA: Sage, 1984).

32. L. Wechsler, "The Great Exception, Part I: Liberty," *The New Yorker*, April 3, 1989, 43–85; "Part II: Impunity," *The New Yorker*, April 19, 1989, 85–108.

33. Judith L. Herman, *Trauma and Recovery* (New York: Basic Books, 1982).

8

An Elephant in the Room

The Impact of Traumatic Stress on Individuals and Groups

Sandra L. Bloom, M.D.

Introduction

If you are an American, you have a greater than fifty-fifty chance of experiencing an event that is commonly recognized as traumatic at sometime in your lifetime and a substantial minority of you will experience three or more traumatic events. Of those of you who do sustain a trauma, at least 25 percent will go on to develop post-traumatic stress disorder (PTSD), a combined physical and psychological disorder that remains chronic, severe, and permanent in 40 percent of cases.[1] If you do develop PTSD, you are eight times as likely as anyone else to end up being diagnosed with three or more psychiatric problems.[2] If you are raped or physically assaulted, your likelihood of developing PTSD rises dramatically: depending on the study you read, 50–70 percent of rape victims develop PTSD. This is particularly disturbing because if you are an adult American woman, you stand a one in eight chance of being raped and a fifty-fifty chance of suffering from violence at the hand of an intimate partner.[3] If you are pregnant, you have up to a one in four chance of being battered and over thirty-two thousand pregnancies each year are attributable to rape.[4] If you are an adolescent girl, you have a one in five chance of being battered or raped by someone you are dating.[5] If you happen to suffer from a severe mental illness, you have a 97 percent chance of being violently victimized at some point in your life.[6] And these numbers just cover the impact of adult exposure to trauma. There is insufficient space in this chapter to explore the enormously devastating impact of child abuse and neglect but suffice it to say that the more categories of childhood adverse events you have sustained, the greater the likelihood that you will suffer from severe substance abuse problems, will try to commit suicide, and will have a greater likelihood of suffering from heart disease, cancer, liver disease, skeletal fractures, and lung disease.[7]

If over half of the population are at such risk for so many significant health and mental health problems—and here I have only touched the surface of the extensive body of research that substantiates the connection between trauma exposure and a panoply of serious and long-term, sometimes lethal problems—it is odd that so little attention has been paid to the role traumatic experience plays in our individual and social health to say nothing of the impact on historical and political events as well as institutional organization and ethical systems of belief. One would think that the interconnections between traumatic experience and many other problems would be as obvious as, say, an elephant in the middle of a small room. But even in the field of mental health treatment, where the issue of exposure to trauma would seem to be most apparent and unavoidable, identification of the traumatic origins of many psychiatric problems still struggles for recognition. The serious study of trauma has waxed and waned over the centuries, usually tied to a social movement, but only in the last twenty years has an extensive body of scientific research begun to illuminate the many ways in which stress, particularly overwhelming stress, affects the minds, bodies, and meaning schemas of individuals.[8]

At this point we can only speculate about what meaning can be derived from generalizing the findings about individuals to larger social contexts, but given the high incidence of trauma in the population, it would appear to be a potentially fruitful field of study. In our evolutionary past, human survival depended on a number of adaptations evolved from earlier species that offer key insights about the ways we continue to be influenced by our evolutionary past, even while social evolution radically alters the environment to which we must now adapt. It may not be an exaggeration to assert that the traits and abilities that guaranteed our species survival in the evolutionary past now threaten our continued individual and collective survival in the present.[9] To explore further how these ideas may help us understand the aftereffects of September 11, 2001, in this chapter we will look at what happens to individuals and groups when a threat occurs, the ways in which the original coping responses to stress become disabling problems under the influence of recurrent threat, and the implications of what we now know about the psychobiology of stress for individual and social healing.[10]

It is my premise that parallel processes are at work in all human systems and they can stand in as metaphors, if not actual representations, for each other.[11] The result of the parallel process nature of human systems is that our organizations and society as a whole frequently recapitulate for individuals the very experiences that have proven so toxic for them, while individual reenactment tends to shape the structure and function of those institutions. This complex, multigenerational interaction can produce ever-worsening dysfunction in both individuals and systems. Since the disasters that occurred on September 11, 2001, interest has increased in a framework that can adequately describe the impact of collective trauma and collective disaster.[12]

Given the actual incidence of exposure to trauma and the negative impact of multiple traumatic experiences, viewing this as purely an individual problem is not enough. It is critical that we design cultures that are less traumatizing and that offer more opportunities for individuals and families to recover from exposure to violence. It is possible that simultaneous individual and institutional change could redirect the course of social evolution in a less destructive direction. In order to promote social evolution rather than devolution, we require a different framework within which to think about the problems that confront us, a framework that enables us to strategize alternative forms of action that control the same biological drives that if not checked may lead us over the precipice to destruction. You cannot begin to solve a problem unless you have correctly identified exactly what the problem is. It is the premise of this chapter that we have not correctly identified the problem and that we are still struggling to see the full shape and size of the elephant in the room.

When Terror Strikes

Unlike other mammals, we come into the world ill-prepared to do battle with the natural enemies that surrounded us in our evolutionary past. Helpless for a prolonged period after birth, bearing fragile bodies that lack substantial protection, we have few natural defenses. Like all mammals, we are equipped to respond to emergencies with what is called the "fight-flight-freeze" reaction, also known as the "human stress response."[13] The stress response is a total body-mind mobilization of resources. Powerful neurochemicals flood our brain and body. Our attention becomes riveted on the potential threat and our capacity for reasoning and exercising judgment is negatively impacted by the rising anxiety and fear. This state of extreme hyperarousal serves a protective function during an emergency, preparing us to respond rapidly to any perceived threat, preferentially steering us toward action and away from the time-consuming effort of thought and language. Taking action appears to be the only solution to this extraordinary experience of tension, so we are compelled to act on impulses that often direct us to aggressively defend ourselves rather than to submit or run away.

Our method for remembering things, processing new memories, and accessing old memories is radically changed when under stress. A growing body of evidence indicates that there are actually two different memory systems in the brain—one for verbal learning and remembering that is based on words, and another that is largely nonverbal.[14] The memory we consider our "normal" memory is a system based on language. Under normal conditions, the two kinds of memory function in an integrated way. Our verbal and nonverbal memories are thus usually intertwined and complexly interrelated. However, the human verbally based memory system is particularly vulnerable to high levels of stress. Like our animal ancestors who lacked verbal communication, we become less attentive to words and far more focused on threat-related signals in the environment—all of the nonverbal content of communication. As fear rises, we may lose language functions altogether, possibly mediated by the effect of rising levels of cortisol on the language centers of the brain.[15]

Without words, the mind shifts to a mode of cognition characterized by visual, auditory, olfactory, and kinesthetic images, physical sensations, and strong emotions. This system of processing information is adequate under conditions of danger because it is a more rapid method for assimilating information. By quickly providing data about the circumstances surrounding the danger and making rapid comparisons to previous experience, people may have a vastly increased possibility of survival in the face of threat.

However, there is a problematic side to this emergency adaptation. When the capacity to encode information in language is radically altered under severe stress and the person experiences "speechless terror," the result may be amnesia for the traumatic event—the memory is there, but there are no words attached to it, so it can be neither talked about nor thought about, though it may be expressed through behavior and physical symptoms.[16]

As the level of arousal increases "dissociation"—the loss of integrated function of memory, sensation, perception, and identity—may be triggered as an adaptive response to this hyperaroused state, physiologically buffering the central nervous system and the body by lowering heart rate and reducing anxiety and pain. This internal state of "freeze" helps temporarily reduce the overwhelming nature of the stress response and allows us to stay calm and function rather than experience emotions that are more than we can bear.[17] When overly stressed, human beings cannot think clearly, nor can we consider the long-term consequences of behavior. It is impossible to weigh all of the possible options before making a decision or to take the time to obtain all the necessary information that goes into making good decisions. Decisions tend to be based on impulse and immediate consequences without consideration to unintended consequences. As a result, such decisions tend to be inflexible, oversimplified, extremist, directed toward action, and often very poorly constructed. But although our cognitive function may be oriented entirely toward the present emergency, our associational brain guarantees that we can make hundreds, even thousands, of associations to any event, and the more dangerous the event, the more likely that we will make a multitude of interconnected associations. Later, traumatic memories may be triggered by any reminder of the previous event.[18] Like a complex spider web, a seemingly distant connection can trigger a rapid network of associations that culminates in a physical and sensory experience that is called a "flashback."

The Threatened Group

In our evolutionary past, the development of extended social networks increased the likelihood that vulnerable offspring would be protected and, in combination with our expanding intelligence, made hunting and food gathering far more successful. Human beings could accomplish much more in groups than any one individual could on his or her own. Part of the evolved response to stress that built on our capacity for attachment was a strong inclination to gather together in groups whenever threatened.[19] Under severe stress, emotional arousal becomes so intense that if emotional responses are not buffered by others through social contact and physical touch, our central nervous system is left exposed to unrelenting overstimulation. The result can be long-lasting harm to our bodies as well as our psyches. Our capacity to manage overwhelming emotional states is shaped by our experience with early childhood attachments and is maintained throughout life via our attachment relationships.

Under threat, human beings will more closely bond together with their identified group, close ranks, and prepare for the defense of the group. Human groups under stress tend to become less democratic and more hierarchical and authoritarian, a group structure that lends itself to rapid response. But this rapidity of response sacrifices more complex group processing of information that is typical of democratic interactions. A leader rapidly emerges within such a group, a complex process that is an interaction between the individual characteristics of the leader, the needs of the group, and the contextual demands of the moment. Under such conditions, the vast majority of human beings become more suggestible to the influence of a persuasive, strong, assertive, and apparently confident leader who promises the best defense of the group, thereby containing the overwhelming anxiety of every member of the group. A leader who is attempting to be thoughtful and cautious may instead be considered equivocating or weak when he or she fails to adequately channel the group's anxiety by taking action. Likewise, a leader who drives the group toward action, regardless how ill-considered, may be lauded as strong, noble, and courageous.

Decisions are made quickly and the process of decision making is characterized by extremist thinking, a deterioration of complex processes into oversimplified, dichotomous choices. Decisions are often made autonomously by the leader with relatively little input and the input that he receives is likely to be significantly colored by the pressure everyone feels to conform to standards of group cohesion and unanimity. As stress increases, the leader is compelled to take action to reduce the threat while the followers simultaneously become more obedient to the leader in order to ensure coordinated group effort.

The development of human moral reasoning and our desire for justice can be recognized in early evolutionary development as well. Social relationships are built on the logic of reciprocity, or "tit for tat," probably the basis of all cooperative relationships.[20] Out of betrayed reciprocal relationships comes the natural desire for retaliation or revenge. Out of this innate desire for revenge comes our need to achieve satisfaction for injury and eventually our uniquely human system of laws designed in part to contain and channel vengeance.[21] Under stress, in-group cohesion and territoriality increase and the desire for retaliation for real or imagined violations is increased. Internal conflict and dissent from the group opinion is actively suppressed while out-group projection of hostility onto an external enemy is encouraged, combining to produce a cohesive and organized group that is prepared to fight whoever is designated as the enemy.

Long-standing interpersonal conflicts within the group seem to evaporate and everyone pulls together toward the common goal of group survival, producing an exhilarating and even intoxicating state of unity, oneness and a willingness to sacrifice one's own well-being for the sake of the group. This is a survival strategy ensuring that in a state of crisis decisions can be made quickly and efficiently thus better ensuring survival of the group, even while individuals may be sacrificed.

Threatened Meaning

The development of language was a profound leap forward for the human species. The spoken and later the written word enabled us to share information so that something learned by one individual could be easily and rapidly dispersed among the entire group. Through language, learning could be transmitted not only over space, but over time, so that the knowledge of one generation could be passed on to the next. As our memory system became increasingly more complex we developed two integrated forms of memory, one based on words, the other on nonverbal experience derived from our bodies and our senses. Over time, in fact, we became more and more word-dependent, ultimately basing our sense of reality, our sense of time, and even our sense of self on our word-based intelligence and shared memory, often minimizing or even excluding the importance of nonverbal intelligence, relegated to the largely disrespected sphere of "intuition" or "the arts."

As emotions, intelligence, relational capacity, language, and memory became more fully integrated and as we could compare contemporary experience with the wisdom of the previous generations while anticipating the future, we became desperately aware of our own mortality, a realization so overpowering and awesome that it demanded the creation of meaning systems that could serve to buffer our vulnerable central nervous system against the terror inspired by the mystery of death. Mythology, religion, and philosophy all reflect this meaning-making necessity.[22] Through the creation of shared culture, we became able to fend off the terror of inevitable death. We live with protective illusions of invulnerability that are necessary for health but exposure to overwhelming stress shatters these necessary illusions and for a time, at least, the trauma survivor lives within a world of unreality, a place of devastating confusion, anxiety, and loss.[23] Traumatized groups may also deteriorate rapidly when the cultural underpinnings of reality are likewise shattered and this group effect may compound the individual's disorientation as well.

But what happens when the violation of this meaning space is not a result of one car accident, or one rape, or one flood? What happens when this mind-body-spirit wrenching goes on repeatedly? The tragedy that lies behind our magnificent evolutionary success emerges most fully when a human being is repeatedly traumatized, particularly when that exposure begins in childhood. This effect is multiplied when the traumatized individual is living within a traumatized group. Under such conditions, evolutionary survival mechanisms, so adapted to our continued existence, become dangerous threats and impediments to further growth.

When Terror Becomes A Way of Life

If people are exposed to danger repeatedly, their bodies become unusually sensitive so that even minor threats can trigger off this sequence of physical, emotional, and cognitive responses—a state of chronic hyperarousal.[24] Each episode of danger connects to every other episode of danger in their minds, so that the more danger they are exposed to, the more sensitive they are to danger.[25] With each experience of fight-flight-freeze, their mind forms a network of connections that is triggered with every new threatening experience.

When hyperarousal stops being a state and turns into a trait, human beings lose their capacity to assess and predict danger accurately, leading to avoidance and reenactment instead of adaptation and survival.[26] Prolonged hyperarousal can have disastrous physical effects as biological systems become progressively exhausted. This hyperaroused state makes it likely that people will seek out any substance—drugs, alcohol, food—or behavior that helps provide relief, calm, or distraction.

Childhood exposure to trauma has even more dire consequences than when an adult experiences a traumatic event for the first time. Children's brains are still forming. The release of powerful neurohormones, particularly during critical and sensitive moments in development, is thought to have such a profound impact on the developing brain that the brain may organize itself around the traumatic event. We are only beginning to understand how the effects of chronic stress set the stage for long-term physical as well as emotional and social problems.[27]

The experience of overwhelming terror destabilizes the internal system that regulates emotional arousal. Usually, people respond to a stimulus based on the level of threat that the stimulus represents. People who have been traumatized lose this capacity to modulate arousal and manage affect and this loss of control can negatively impact on a number of important functions. Emotional management is critical to learning and the capacity to exercise reasoned judgment. Emotions prioritize thinking by directing attention to important information. Emotions are sufficiently vivid and available to be used as aids for judgment and memory. Emotional mood swings change one's perspective, encouraging multiple points of view, and emotional states differentially encourage specific problem approaches.[28]

Children who are exposed to repeated episodes of overwhelming arousal do not have the kind of safety and protection that they need for normal brain development and therefore they may never develop normal modulation of arousal and this severely compromises their capacity for emotional management. As a result they are chronically irritable, angry, impulsive, anxious, and unable to manage aggression. The compromised emotional management interferes with learning and the development of mature thought processes.[29]

This failure to develop healthy ways of managing emotional arousal also interferes with relationships. Mature emotional management endows us with the abilities to interpret the meanings that emotions convey regarding relationships, to understand complex feelings, and to recognize likely transitions among emotions. The gradual acquisition of this emotional intelligence allows us to monitor emotions in relation to ourselves and others while giving us the ability to manage emotion in ourselves and others by moderating negative emotions and enhancing pleasant ones without repressing or exaggerating the information they convey.

Children—and the adults they become—who experience compromised emotional management are likely to experience high levels of anxiety when alone and/or in interpersonal interactions. Therefore, they will understandably do anything they can to establish some level of self-soothing and self-control. Under such circumstances, people frequently turn to substances, such as drugs or alcohol, or behaviors such as sex or eating or risk-taking behavior, or even engagement in violence, including self-mutilation, all of which help them to calm down, at least temporarily, largely because of the internal chemical effects of the substance or behavior.[30] Human beings, human touch, could also serve as a self-soothing device, but for trauma survivors, trusting human beings may be too difficult.

Compromised emotional management skills are also one mechanism of intergenerational transmission since these skills build up over time in the interaction between parent and child. A parent who has compromised skills will be unable to provide the important emotional learning experiences that their children require. Instead, the children will adapt to the parental style of managing emotions.

Human beings deplore being helpless. Placed into situations of helplessness, we will do anything to escape and restore a sense of mastery. But helplessness is a hallmark characteristic of a traumatic experience. Helplessness in the face of danger threatens survival and our carefully established sense of invulnerability and safety. Worse yet, repetitive exposure to helplessness is so toxic to emotional and physiological stability that in service of continued survival, survivors are compelled to adapt to the helplessness itself, a phenomenon that has been termed "learned helplessness."[31] Like animals in a cage, with enough exposure to helplessness people will adapt to adversity and cease struggling to escape from the situation, thus conserving vital resources and buffering the vulnerable central nervous system against the negative impact of constant overstimulation. Then, rather than change situations that could be altered for the better, they will change their definitions of "normal" to fit the situation to which they have become adapted. Later, even when change is possible, the formerly adaptive response of simply buckling down and coping can create a serious obstacle to positive change, empowerment, and mastery. This may contribute to the dynamic of revictimization.[32]

As a result of this adjustment, people who have had repeated experiences of helplessness will exhibit a number of apparently contradictory behaviors. On the one hand they may demonstrate "control issues" by trying to control other people, themselves, their own feelings—anything that makes them feel less helpless. The threat of or use of violence is a way to control other people that is frequently effective. At the same time, they are likely to be willing to turn over control to substances or behaviors that are destructive, or to people who cannot be trusted. They may have difficulties discriminating between abusive and healthy authority and may be willing to give up control to abusive authorities. Springing from similar experiences and dynamics, one victim goes on to become a perpetrator, another to become revictimized, another becomes neither, all reflecting multiple and complex decision paths that the person begins walking very early in life.

The adjustment to adversity also keeps them from making positive changes when they could do so. Once a human being has adjusted to adverse conditions, these conditions are accepted as "normal" and any change from what feels normal is resisted. Changed conditions become a habit. We are basically conservative creatures and we resist changing habits once we have developed them and the more associated the habit formation has been with danger and surviving a threat, the less likely we are to change it and the more likely we are to resist attempts to get us to change. Instead, we unknowingly shift our internal norms. Once we have reset our norms, we tend to repeat the past simply because it seems oddly comfortable to do so, even when cognitively we appraise the situation as being less than ideal. When the past is a traumatic one, then we are likely to be victimized again and again in a progressively downward spiral, while we internally believe that there is really nothing we can do about it—it's just the way things are.

Our very complex brains and powerful memories distinguish us as the most intelligent of all animals, and yet it is this very intelligence that leaves us vulnerable to the effects of trauma such as flashbacks, body memories, posttraumatic nightmares, and behavioral reenactments. Exposure to trauma alters people's memory, producing extremes of remembering too much and recalling too little. Unlike other memories, traumatic memories appear to become etched in the mind, unaltered by the passage of time or by subsequent experience.[33] But without verbal content, traumatic memories are not integrated into the narrative stream of consciousness but instead remain as unintegrated fragments of experience that can then intrude into consciousness when triggered by reminder of a previous event.

A flashback is a sudden intrusion of a fragment of past experience into present consciousness. A flashback may take the form of a visual image, a smell, a taste, or some other physical sensation including severe pain, and is usually accompanied by powerful and noxious emotions. Even thinking of flashbacks as "memories" is inaccurate and misleading. When someone experiences a flashback, he does not remember the experience, but relives it. Often the flashback is forgotten as quickly as it happens because the two memory systems are so disconnected from each other. Every time a flashback occurs, the complex sequence of psychobiological events that characterize the fight-flight-freeze response is triggered, resulting in a terror reaction to the memories themselves. The result is a vicious cycle of flashback-hyperarousal-dissociation that further compromises function. As the survivor tries to cope with this radical departure from normal experience, he or she will do anything to interrupt the vicious cycle—drugs, alcohol, violence, eating, sex, risk-taking behaviors, self-mutilation—all can temporarily produce an interruption. But each in its own way compounds the individual's growing problems.

Our dependence on language means that wordless experiences cannot be integrated into consciousness and a coherent sense of identity, nor will those experiences rest quietly. Instead, the survivor becomes haunted by an unnarrated past. Since the sense of "self" refers to a verbally based identity, experiences that have not been encoded in words are not recognized as a part of the "self." Lacking an ability to talk to themselves—an internal dialogue that is going on all the time—controlling impulses is exceedingly difficult.

"Traumatic reenactment" is the term used to describe the lingering enactment and automatic repetition of the past. The very nature of traumatic information processing determines the reenactment behavior. As human beings, we are physically designed to function at a maximum level of integration and any barrier to this integration seems to produce some innate compensatory mechanism that potentially allows us to overcome it. Based on what we know about the split between verbal and nonverbal thought, it may be that the most useful way of understanding traumatic reenactment is through the language of drama. For healing to occur, victims must give words and meaning to their overwhelming experiences. The traumatized person is cut off from language, deprived of the power of words, trapped in speechless terror. The only way that the nonverbal brain can "speak" is through behaviors. This is the language of symptoms, of pathology, of deviant behavior in all its forms. Unfortunately, we have largely lost the capacity for nonverbal interpretation, and we have ceased to take the time to examine and understand repetitive patterns of behavior. As a result, most of these symptomatic "cries for help" fall on deaf ears. Instead, the society judges, condemns, excludes, and alienates the person who is behaving in an asocial, self-destructive, or antisocial way without hearing the meaning in the message. Trapped in a room with no exit signs, they hunker down and adapt to ever-worsening conditions, unaware that there are many opportunities for change and terrified that taking any risk to get out of their dilemma could lead to something even worse.

Exposure to chronic severe stress may disrupt the attachment system so that stress—instead of social support—is associated with anxiety relief, an outcome known as "addiction to trauma," further damaging the attachment system and creating an increased likelihood that people will turn to self-destructive behavior, addictive substances, violence, and thrill-seeking as a way of regulating their internal environments.[34] Even more ominous for repeatedly traumatized people is their pronounced tendency to use highly abnormal and dangerous relationships as their normative idea of what relationships are supposed to be.[35] "Trauma-bonding" describes a relationship based on terror and the twisting of normal attachment behavior into something perverse and cruel.[36] Relationships to authority also become damaged. Human beings first learn about power relationships in the context of the family. If we experience fair, kind, and consistent authority figures, then we will internalize that relationship to authority both in the way we exert control over our own impulses and in the way we deal with other people. If we have been exposed to harsh, punitive, abusive, inconsistent authority then the style of authority that we adopt is likely to be similarly abusive.

The human ability to form healthy attachments to other people allows us to transit successfully through the process of grieving after a loss. People who have disrupted attachment experiences have difficulties with grieving. New losses tend to open up old wounds that never heal. Arrested grief is extremely problematic because it is impossible to form healthy new attachments without first finishing with old attachments. In this way, unresolved loss becomes another dynamic that keeps an individual stuck in time, unable to move ahead, unable to go back. Compounded and unresolved grief is frequently in the background of lives based on traumatic reenactment.[37]

As this process of prolonged hyperarousal, helplessness, emotional numbing, disrupted attachment, and reenactment unfolds, people's sense of who they are, how they fit into the world, how they relate to other people, and what the point of it all is, can become significantly limited in scope. As this occurs, they are likely to become increasingly depressed. The attempt to avoid any reminders of the previous events, along with the intrusive symptoms, such as flashbacks and nightmares, comprise two of the interacting and escalating aspects of post-traumatic stress syndrome, set in the context of a more generalized physical hyperarousal. As these alternating symptoms come to dominate traumatized people's lives, they feel more and more alienated from everything that gives their lives meaning—favorite activities, other people, a sense of direction and purpose, a sense of spirituality, a sense of community. It is not surprising, then, that slow self-destruction through addictions, or fast self-destruction through suicide, may be the final outcome of these syndromes. For other people, rage at others comes to dominate the picture and these are the ones who end up becoming significant threats to other people as well as themselves.

Children who are traumatized do not have developed coping skills or a developed sense of self or self in relation to others. Their schemas for meaning, hope, faith, and purpose are not yet fully formed. They are in the process of developing a sense of right and wrong, of mercy balanced against justice. All of their cognitive processes, such as their ability to make decisions, their problem-solving capacities, and learning skills, are still being acquired. As a consequence, the responses to trauma are amplified because they interfere with the processes of normal development. Living in a system of contradictory and hypocritical values impairs the development of conscience, of a faith in justice, of a belief in the pursuit of truth.

We are meaning-making animals. We must be able to make sense of our experience, to order chaos and structure our reality. Traumatic experience robs people of a sense of meaning and purpose. It shatters basic assumptions about the nature of life and reality.[38] Close contact with traumatic death or threats to our own mortality cannot be accepted but can only be transcended and trauma dramatically interferes with our capacity to grow, to change, and to move on. Shared cultural beliefs largely determine the way human beings cope with the terror of inevitable death, and traumatic experiences disrupt the individual's sense of individual and cultural identity.[39] Losing the capacity for psychic movement, they deteriorate into a repetitive cycle of reenactment, stagnation, and despair.

The end result of this complex sequence of posttraumatic events is repetition, stagnation, rigidity, and a fear of change all in the context of a deteriorating life. As emotional, physical, and social symptoms of distress pile on each other, victims try desperately to extricate themselves by using the same protective devices that they used to cope with threat in the first place—dissociation, avoidance, aggression, destructive attachments, damaging behaviors, and addictive substances. The response to threat has become so ingrained and automatic that victims experience control as beyond them, and as their lives deteriorate, their responses become increasingly stereotyped and rigid.

Recurring Threats in Group Life

Group responses to stress are measures that may be extremely effective during an acute state of crisis. However, chronic and recurring threats to the group can lead to states as dangerous for a family, an organization, or a nation as chronic hyperarousal is to the health and well-being of the individual. When a group atmosphere becomes one of constant crisis, with little opportunity for recuperation before another crisis manifests, the toxic nature of this atmosphere tends to produce a generally increased level of tension, irritability, short tempers, and even abusive behavior.

The urgency to act in order to relieve this tension compromises decision making because group members are unable to weigh and balance multiple options, arrive at compromises, and consider long-term consequences of their actions under stress. Decision making in such groups tends to deteriorate over time with increased numbers of poor and impulsive decisions, compromised problem-solving mechanisms, and overly rigid, extremist, and dichotomous thinking and behavior. Interpersonal conflicts that were suppressed during the initial crisis return, often with a vengeance, but conflict resolution mechanisms, if ever in place, deteriorate further under the influence of chronic stress.

Unable to engage in complex decision making, group problem solving is compromised, making it more likely that the group will turn to—or continue to support—leaders who appear strong and decisive, and who urge repetitive but immediate action that temporarily relieves tension and may even bring a sense of exhilaration. Leaders may become increasingly autocratic, bullying, deceitful, and dogmatic, trying to appear calm and assured in front of their followers while narrowing their circle of input to a very small group of trusted associates. As the leader becomes increasingly threatened, sensing the insecurity of his decisions and his position, these small groups of associates feel increasingly pressured to conform to whatever the boss wants and are more likely to engage in groupthink.[40] In this process, judgment and diversity of opinion are sacrificed in service of group cohesion and, as this occurs, the quality of decision making becomes compromised, progressively and geometrically compounding existing problems.

Escalating control measures are used to repress any dissent that is felt to be dangerous to the unity of what has become focused group purpose. This encourages a narrowing of input from the world outside the group. Research has demonstrated that when threatened by death, people will more strongly support their existing cultural belief systems and actively punish those who question those belief systems.[41] So, any subgroups that attempt to protect against unfolding events are harshly punished.

If group cohesion begins to wane, leaders may experience the relaxing of control measures as a threat to group purpose and safety. They may therefore attempt to mobilize increasing projection onto the designated external enemy who serves a useful purpose in activating increased group cohesion while more strenuously suppressing dissent internally. This cycle may lead to a state of chronic repetitive conflict externally and escalating repressive measures internally.

This entire process tends to increase instead of decrease the sense of fear and insecurity on the part of everyone within the group, and as leaders focus exclusively on physical security, the group may be willing to sacrifice other forms of safety and well-being in order to achieve an elusive sense of physical security that remains threatened. In doing so the group may endorse rapid changes that result in the creation of new rules without considering the unintended consequences that may include the widespread loss of rights, liberty, and freedom. In order to restore the illusion of safety that can then only be secured through an endless escalation of hostility, aggression, and defense, projection onto an external enemy, suppression of internal conflict, and demands for greater group loyalty all intensify. Group norms shift radically but insidiously, the changes cloaked within a fog of what appears to be rational decision making.

As time goes on and the recurrent stress continues, the group will adapt to adversity by accepting changed group norms. Although this adjustment to changed group norms feels normal, actual behavior becomes increasingly aberrant and ineffective. When someone mentions the fact of the changed norms, about the differences between the way things are now and the way they used to be (when the group was more functional), the speaker is likely to be silenced or ignored. As a result there is an escalating level of acceptance of increasingly aberrant behavior.

Like individuals, groups can forget their past and the more traumatic and conflicted the past the more likely it is that groups will push memories out of conscious awareness. Critical events and group failure change us and change our groups, but without memory we lose the context. As in families, so too in societies, past traumas are frequently known and not known—historically recognized but never really talked about, mourned, or resolved. This is particularly true when the past traumas are also associated with guilt. Studies have shown that institutions do have memory and that once interaction patterns have been disrupted these patterns can be transmitted through a group so that one "generation" unconsciously passes on to the next norms that alter the system and every member of the system. But without a conscious memory of events also being passed on, group members in the present cannot make adequate judgments about whether the strategy, policy, or norm is still appropriate and useful in the present.[42] This process can present an extraordinary resistance to healthy group change.

Groups that have experienced repeated stress and traumatic loss can also experience disrupted attachment. In such a system there will be a devaluation of the importance of relationships. People are treated as widgets, replaceable parts that have no significant individual identity or value. There is a lack of concern with the well-being of others as the group norm, perhaps under the guise of "don't take it personally—it's just business." In groups with disrupted attachment schemata there is a high frequency of acceptance or even active encouragement of addictive behavior including substance abuse. There is also an unwillingness and inability to work through loss so that people leaving the group are dealt with summarily and never mentioned again. The result is that the group becomes more stagnant and disconnected from a meaningful environment, group loyalty plummets, and productivity declines.

A group that cannot change, like an individual, will develop patterns of reenactment, repeating the past strategies over and over without recognizing that these strategies are no longer effective. With every repetition there is instead further deterioration in functioning. Healthier and potentially healing individuals may enter the group but are rapidly extruded as they fail to adjust to the reenactment role that is being demanded of them. Less autonomous individuals may also enter the group and are drawn into the reenactment pattern. In this way, one autocratic and abusive leader leaves only to be succeeded by another.

When guilt is involved, it is common to find projection and displacement of unpleasant realities onto an external enemy. The continued use of projection over time causes increased internal group splitting and a loss of social integration. Absent a language that engages feeling and the multiple narratives of history, a group cannot heal from past traumatic events and is therefore compelled in overt or symbolized ways to repeat those events.

Similar to a chronically stressed individual, as group stress persists, a pattern of group failure begins to emerge. Unable to deal with the increasing complexities of an ever-changing world because of the rigidity and stagnation of problem solving and decision making, the group looks, feels, and acts angry, depressed, and anxious, but helpless to effect any change. There is an increasing rate of illness, addiction, and antisocial acts among the individuals within the group. Burnout, personality distortions, and acting-out behavior all increase. Conflicts arise repeatedly and are not resolved or even addressed. As this deterioration continues, group members feel increasingly demoralized and hopeless, concerned that the group mission and value system has been betrayed in countless ways.

Alienation begins to characterize the social milieu and evidence for it can be seen in increased internal splitting and dissension, rampant hypocrisy, a loss of mutual respect and tolerance, apathy, cynicism, hopelessness, helplessness, loss of social cohesiveness and purpose, loss of a sense of shared social responsibility for the more unfortunate members of any population, and the loss of a shared moral compass. Alienation is the end result of an unwillingness or inability to work through the fragmentation, dissociation, and disrupted attachment attendant on repetitive traumatic experience. Increasing feelings of alienation are symptoms of severe degradation and stagnation and signal that the time for systemic change is at hand if the organism is to survive. System-wide corruption, systematic deceit, empathic failures, abusive laws, increasingly punitive laws, hypermoralism, hypocrisy, and a preoccupation and glorification of violence are all symptomatic of impending group bankruptcy or system failure.

In this parallel process way, human systems can inadvertently recapitulate the very experiences that have proven to be so toxic for the individuals who populate those very groups. Groups designed to help people survive more successfully may end up becoming "trauma-organized systems," inadvertently organized around interactively repeating the patterns of repetition that are keeping the individuals they are serving from learning, growing, and changing. The inefficient or inadequate fulfillment of the mission of the group—since all groups have as their evolutionary mission mutual protection—takes a toll on individuals within the group and wastes money and resources. This vicious cycle also lends itself to a worldview that the most injured people in any group are the cause of the problem and that their situations are hopeless and they cannot really be helped and the failure of group support leads to a self-fulfilling prophecy as the injured individuals continue to deteriorate.

Individual Recovery from the Effects of Repetitive Trauma

Recovery from even severe and repetitive exposure to trauma is possible but the road to recovery is challenging and variable and may be remarkably complex. The reasons for this complexity lie in the extent and many-layered aspects of the injuries that people sustain. Similar to physical healing, psychological healing requires safety and protection from further injury. Healing from trauma requires enormous psychological courage and endurance. Like lancing an abscess, the toxic memories and emotions must be drained—or as a nineteenth-century pioneer, Pierre Janet, described it, "liquidated."[43]

This is a painful process that human beings naturally resist. For change to occur, there has to be enough "heat" to move the person out of the stable, but unhealthy, equilibrium state at which they have arrived. Much of what occurs in life and in treatment requires movement into "far-from-equilibrium" conditions as the chaos theorists call it.[44] Recovery requires that people do things that do not feel natural or comfortable at all but that do destabilize equilibrium—endure pain, change habits, face unpleasant memories, end destructive relationships, initiate healthier relationships, accept vulnerability, tolerate the breakdown of meaning, grieve for losses, try new things. Critical to the willingness to engage in this process is an emerging vision of something better, a life after trauma, where the past no longer haunts the present. This is the new "attractor" that draws the survivor toward it, allowing him or her to move into a new equilibrium state. Given all this, it is not surprising that many people do not heal—it is more surprising that so many do.[45]

The first step in healing is to achieve a higher level of safety. This requires recognition of the nature and extent of the injuries that people have sustained: injuries to the ability to achieve basic physical safety; injuries to the ability to be safe with oneself—psychological safety; injuries to the ability to be safe with others—social safety; and freedom from harm within a shared value system—ethical safety. The achievement of safety is relational in that the traumatized individual must be willing to commit to the active pursuit of nonviolence while the social group must be willing to supply whatever resources the individual requires to reduce threat, minimize physiological hyperarousal, attend to physical illness, and learn affect management and cognitive-behavioral skills necessary to become safe.[46]

Traumatized people need to develop self-soothing techniques and new skills for managing overwhelming affect without losing control and acting out destructively. They must learn how not to dissociate under stress. They are likely to require the enhanced development of better cognitive skills to improve decision making and problem solving as well as methods for resolving conflicts without resorting to violence. They must learn to alter their attitudes toward authority figures and other people in their lives and this requires learning to trust other people—a difficult proposition under the best of circumstances. The tendency to unconsciously reenact the dynamics of a previous traumatic experience in current relationships must be skillfully and compassionately confronted and redirected. This may be profoundly difficult without specific trauma-resolution techniques that offer important but painful opportunities to integrate nonverbal fragments of past-life experience into verbal narrative form, enabling the past to be placed back within a proper time sequence and context. Recovery of the past necessitates engaging in the process of grieving for all that has been lost, otherwise arrested grief will prevent forward movement and the restoration of the capacity for healthier relationships. Merging into and emerging out of this complex process of healing is the continuous search for meaning, the development of a postinjury values system, embedded within a restored cultural context.

The least understood—but arguably the most important ingredient for recovery from severe trauma—is vision. Without hope that life can be improved, that wounds can be healed, that obstacles can be overcome, a human being is unable to endure the painful work of resolving traumatic loss. Engendering hope is also relational, an existential leap of faith between survivors, the groups within which their lives are embedded, the people who went before them, those who come after them, and whatever version of a Higher Power inspires faith in them that there is some purpose beyond what can be immediately seen.

Group Recovery from the Effects of Repetitive Trauma

Healing from recurrent trauma is not an exclusively individual process. Just as traumatic injury occurs within a social context so too does recovery from those injuries. Because recovery is so complex and healing needs to occur on so many levels simultaneously and interactively, there is an interdependent relationship between individual and culture.[47] The more damaged the individual is, the more likely he or she is to contribute negatively to traumatic dynamics within the culture. Likewise, the more traumatized the culture is, the more difficult it will be for the traumatized individual to find the resources necessary to pursue recovery.[48]

Spared throughout most of our history of significant, externally derived national trauma, the events of September 11, 2001, permanently shattered a myth of invulnerability expressed in the often-heard phrase "America will never be the same again." In reality, we were in as much danger on September 10 as we were on September 12, but the shared cultural belief that death would have no dominion here, was lost. Beginning on September 11, 2001, the nation entered a state of repetitive alarm, fueled by the color-coded warning system that officials periodically used to frighten the population without providing any accompanying system of mastery that could assist people in successfully managing recurrent experiences of helplessness. The loss of a sense of security and the recurrent sense of helplessness was intolerable. Quickly, national mourning gave way to the drums of war and the grieving process was prematurely arrested and redirected in service of aggression. An intense desire for vengeance fueled the search for an external enemy, and when the real enemy could not be found, an available enemy was substituted. A president who had low ratings before September 11, soared in popularity when he seized the opportunity to take action and satisfy the lust for blood revenge. He was rapidly celebrated as an icon of virtue who could not be criticized or questioned. The nation became severely divided between those who rated his leadership abilities on par with the greatest historical and military leaders and those who believed that in what was described as recurrent deceitful words and actions he led the country into an unnecessary and protracted war and could be considered the most dangerous president the nation has ever had.[49] Meanwhile, a White House that already had a reputation for secrecy became even more guarded about information, now under the guise of national security. Insiders said that there was little open and complex discussion, and diverse points of view were not encouraged.[50] Dissent was voluntarily stifled by the media, by government, and by regular citizens.[51] Safety was defined only in physical terms so that liberty was increasingly sacrificed in the name of security, while all attempts to discuss broader concepts of safety by exploring multicausal aspects of terrorism and terrorist attacks was forbidden, suppressed as unpatriotic and a threat to group cohesion.[52] Protest was forbidden and protestors punished. Wide-ranging laws were enacted with very little discussion about the long-term and potentially very negative consequences of the laws. Decisions were made that on closer scrutiny were premature and poorly planned, particularly about going to war with Iraq and anticipation of the consequences of such a war.[53] Relationships with long-standing allies were broken and actions were taken that antagonized most of the rest of the world. Long-standing unresolved conflicts between parties, between races, between classes, between regions, between minority groups, and between religions all intensified and became more vituperative. This sequence of events describes a large group in the throes of reacting to severe and repetitive stress. As such, it provides an excellent case study for suggesting alternative and potentially more healing methods for groups to recover from traumatic events.

For a group as for an individual, paying attention only to physical safety does not make a group safe. A military response may be necessary. But if it is considered sufficient it may result in a group forced to live within an armed fortress, while true security remains elusive. Real safety depends on at least three other domains: respect for individual rights, a shared sense of social responsibility, and a system of ethical conduct that is expressed as strongly in deeds as in words. Because threats to physical safety rivet our attention, it requires courageous leaders to rein in the group desire for revenge and the longing to take action and to create enough calm to engage the members of the group in the more complex discussions entailed in discussing multiple and interactive layers of true safety. The true antidote to fear is not less democratic discussion but more. Studies indicate that large groups of people may make better decisions than even groups of experts.[54] In times of stress it is even more critical that group leaders solicit diverse points of view, encourage wide-ranging discussion of alternatives, and encourage dissent. It is especially useful if those who have not been immediately traumatized can be party to the discussion because they are likely to be able to convey a more balanced point of view. The wider the discussion, the more likely it is that patterns of reenactment will become visible, that eroding norms will be recognized, and that stress-based responses will be slowed down. Groups must recognize that under conditions of extreme stress, decision making processes are likely to be compromised so decisions that must be made should be short-term and subject to later review. Time must be given to grieve for losses and to work through the attendant fear and anger and pain. Premature action should be discouraged. Most importantly perhaps, leaders must project a vision of hope, a reason for the group members to engage in the difficult work of recovery.

So, given an understanding of the dynamics of the individual and group stress response, what could we have done differently after September 11? The attention paid to enhancing physical security measures was necessary, but it was simply not sufficient. An opportunity was lost to engage an entire social group in broad discussions about the root causes of terrorism in the globally connected world we live in, a discussion that would have been far more complex and difficult but potentially more fertile than the more simple-minded and singular recourse to war. A different leader could have encouraged calm, and discouraged premature action. He—or she—could have allowed sufficient time for mourning and restabilization, while meanwhile pursuing a wide diversity of opinions about how to proceed. Fertile dissent could have been encouraged. Resources could have been devoted to pursuing a strategy to locate and demilitarize the true enemies. The Homeland Security Act could have been more carefully crafted to provide more security without sacrificing our honored social norms of liberty and justice, and time would have allowed it to be actually read by the people that voted on it. We could have had a leader who would have inspired us to hope for something better than a state of chronic fear while not denying the dire nature of the existing problems—a leader who could offer us a vision of what life after trauma could be if we were prepared to make the sacrifices necessary to share in the creation of a better world.

Conclusion

Expanding our understanding of the impact of stress is vitally important. The rapid rate of change, the growth of technology, and the widespread changes associated with globalization, population growth, and the spread of information all contribute to creating stressed social structures around the world. A significant part of the stress that we all face arises from the complexity involved in virtually every situation that confronts us from raising children to global peacemaking. In a globally interconnected world, with diminishing resources, burgeoning populations, and weapons of mass destruction poised on every side, species survival is contingent on our ability to evolve socially more rapidly than can be accommodated by our biological evolution. Faced with the capacity for total species annihilation, the only long-term solution possible is a radical shift in paradigm if we are to survive as a species. Since our physical evolution has left us with one foot in the twenty-first century and the other solidly planted in the Stone Age, this will require a leap in social evolution. To survive, we must deliberately and systematically come to understand the inherent dangers in responding to our present-day enemies as we once did to saber-toothed tigers. War is no longer a viable option because in an interconnected world it is an act of savage self-mutilation. Nonviolent resistance to violence, a more equitable dispersal of wealth, and an increase in democratic processes offer more complex solutions to the exceedingly complex problems that confront us.

So this then is the "elephant in the room"—the current and multigenerational impact of traumatic experience on individuals, families, institutions, and entire societies. As Braudel has noted, "A civilization generally refuses to accept a cultural innovation that calls in question one of its own structural elements. Such refusals or unspoken enmities are relatively rare; but they always point to the heart of a civilization."[55] The information that the study of traumatic experience reveals about the nature of human nature challenges the existing paradigmatic structures that support many of our present social structures. This chapter suggests that we have a great deal to learn from both the posttraumatic development of individual pathology and recovery from trauma-related syndromes. It is my hope that the implications of healing from traumatic events can help guide organizational and social policy efforts to accelerate the process of social evolution and transformation.

Notes

1. M. W. Friedman, *Post Traumatic Stress Disorder: The Latest Assessment and Treatment Strategies* (Kansas City, MO: Compact Clinicals, 2000).

2. R. C. Kessler, A. Sonnega, E. Bromet, M. Hughes, and C. B. Nelson, "Posttraumatic Stress Disorder in the National Comorbidity Survey," *Archives of General Psychiatry* 52, no. 12 (1995): 1048–60.

3. D. G. Kilpatrick, C. Edmunds et al., *Rape in America: A Report to the Nation* (Charleston: Medical University of South Carolina, National Center for Victims of Crime, Crime Victims Research and Treatment Center; and National Victim Center, 1992).

4. National Victim Center, *Crime and Victimization in America: Statistical Overview* (Arlington, VA: National Victim Center, 1993).

5. J. G. Silverman, A. Raj et al., "Dating Violence against Adolescent Girls and Associated Substance Use, Unhealthy Weight Control, Sexual Risk Behavior, Pregnancy, and Suicidality," *Journal of the American Medical Association* 286, no. 5 (2001): 572–79.

6. L. A. Goodman, S. D. Rosenberg et al., "Physical and Sexual Assault History in Women with Serious Mental Illness: Prevalence, Correlates, Treatment, and Future Research Directions," *Schizophrenia Bulletin* 23, no. 4 (1997): 685–96.

7. V. J. Felitti, R. F. Anda et al., "Relationship of Childhood Abuse and Household Dysfunction to Many of the Leading Causes of Death in Adults: The Adverse Childhood Experiences (ACE) Study," *American Journal of Preventive Medicine* 14, no. 4 (1998): 245–58.

8. J. Herman, *Trauma and Recovery* (New York: Basic Books, 1992).

9. B. Ehrenreich, *Blood Rites: Origins and History of the Passions of War* (New York: Henry Holt, 1998).

10. S. L. Bloom, "Neither Liberty nor Safety: The Impact of Fear on Individuals, Institutions, and Societies, Part I," *Psychotherapy and Politics International* 2, no. 2 (2004): 212–28.

11. K. K. Smith and N. Zane, "Organizational Reflection: Parallel Processes at Work in a Dual Consultation," *Journal of Applied Behavioral Science* 35, no. 2 (1999): 145–62.

12. K. Erickson, *A New Species of Trouble: The Human Experience of Modern Disasters* (New York: Norton, 1994.)

13. M. J. Horowitz, ed., *Stress Response Syndromes*, 2nd ed., (Northvale, NJ: Jason Aronson Press, 1986); M. J. Horowitz, *Treatment of Stress Response Syndromes* (Washington, D.C.: American Psychiatric Association Press, 2003).

14. B. A. Van der Kolk, "Trauma and Memory," in *Traumatic Stress: The Effects of Overwhelming Experience on Mind, Body and Society*, ed. B. A. Van der Kolk, A. C. McFarlane, and L. Weisaeth (New York: Guilford Press, 1996), 279–302.

15. B. S. McEwen A. M. Magarinos, "Stress Effects on Morphology and Function of the Hippocampus" in *Psychobiology of Posttraumatic Stress Disorder*, ed. R. Yehuda and A. C. McFarlane (New York: New York Academy of Sciences, 1997) 821:271–84; B. Roozendaal, G. L. Quirarte et al., "Stress-activated Hormonal Systems and the Regulation of Memory Storage," in *Psychobiology of Posttraumatic Stress Disorder*, ed. R. Yehuda and A. C. McFarlane (New York: New York Academy of Sciences, 1997). 821:247–58; B. A. Van der Kolk, "The Complexity of Adaptation to Trauma Self-regulation, Stimulus Discrimination, and Characterological Development," in *Traumatic Stress: The Effects of Overwhelming Experi-*

ence on Mind, Body and Society, ed. B. A. Van der Kolk, A. C. McFarlane, and L. Weisaeth (New York: Guilford Press, 1996), 182–213; Van der Kolk, "Trauma and Memory"; B. A. Van der Kolk, J. A. Burbridge et al., "The Psychobiology of Traumatic Memory: Clinical Implications of Neuroimaging Studies," *Annals New York Academy of Sciences* 821 (1997): 99–113.

16. B. A. Van der Kolk, "The Body Keeps the Score: Memory and the Evolving Psychobiology of Posttraumatic Stress," *Harvard Review of Psychiatry* 1 (1994): 253–65.

17. S. L. Bloom, "Understanding the Impact of Sexual Assault: The Nature of Traumatic Experience," in *Sexual Assault: Victimization across the Lifespan*, ed. A Giardino, E. Datner, and J. Asher (Maryland Heights, MO: GW Medical Publishing, 2003); B. A. Van der Kolk and R. Fisler, "Dissociation and the Fragmentary Nature of Traumatic Memories: Overview," *British Journal of Psychotherapy* 12, no. 3 (1996): 376–90; B. A. Van der Kolk D. Pelcovitz et al., "Dissociation, Somatization, and Affect Dysregulation: The Complexity of Adaptation to Trauma," *American Journal of Psychiatry* 153, no. 7, Festschrift suppl. (1996): 83–93.

18. B. A. Van der Kolk, "Trauma and Memory"; B. A. Van der Kolk, Burbridge et al., "The Psychobiology of Traumatic Memory."

19. D. R. Forsyth, *Group Dynamics*, 2nd ed. (Pacific Grove, CA: Brooks/Cole, 1990).

20. R. Axelrod, *The Evolution of Cooperation* (New York: Basic Books, 1994).

21. S. L. Bloom, ed, *Violence: A Public Health Epidemic and a Public Health Approach* (London: Karnac, 2001).

22. E. Becker, *The Denial of Death* (New York: Free Press 1973); T. Pyszczynski, S. Solomon et al., *In the Wake of 9/11: The Psychology of Terror* (Washington, DC: American Psychological Association, 2003).

23. R. Janoff-Bulman, *Shattered Assumptions: Towards a New Psychology of Trauma* (New York: Free Press, 1992); J. F. Schumaker, *The Corruption of Reality: A Unified Theory of Religion, Hypnosis, and Psychopathology* (Amherst, NY: Prometheus Books, 1995).

24. B. A. Van der Kolk, M. Greenberg et al., "Inescapable Shock, Neurotransmitters, and Addiction to Trauma: Toward a Psychobiology of Post Traumatic Stress," *Biological Psychiatry* 20 (1985): 314–25.

25. J. LeDoux, *The Emotional Brain: The Mysterious Underpinnings of Emotional Life* (New York: Simon and Schuster, 1996).

26. B. A. Van der Kolk, "The Compulsion to Repeat the Trauma: Reenactment, Revictimization, and Masochism," *Psychiatric Clinics of North America* 12 (1989): 389–411.

27. V. J. Felitti, R. F. Anda et al., "Relationship of Childhood Abuse and Household Dysfunction to Many of the Leading Causes of Death in Adults:

The Adverse Childhood Experiences (ACE) Study," *American Journal of Preventive Medicine* 14, no. 4 (1998): 245–58; B. Perry, "Incubated in Terror: Neurodevelopmental Factors in the Cycle of Violence," in *Children, Youth and Violence: Searching for Solutions*, ed. J. Osofsky (New York: Guilford Press, 1995); B. Perry and J. Pate, "Neurodevelopment and the Psychobiological Roots of Post-traumatic Stress Disorder," in *The Neuropsychology of Mental Disorders: A Practical Guide*, ed. L. Koziol and C. Stout (Springfield, IL: Charles C. Thomas, 1994), 81-98; B. D. Perry, R. Pollard et al., "Childhood Trauma, the Neurobiology of Adaptation and 'Use-dependent' Development of the Brain. How 'States' Become 'Traits,'" *Infant Mental Health Journal* 16 (1995): 271–91.

28. J. D. Mayer and P. Salovey, "What Is Emotional Intelligence?" in *Emotional Development and Emotional Intelligence: Educational Implications*, ed. P. Salovey and D. J. Sluyter (New York, Basic Books, 1997), 3–31.

29. B. Perry, "Neurobiological Sequelae of Childhood Trauma: PTSD in Children," in *Catecholamine Function in Posttraumatic Stress Disorders: Emerging Concepts*, ed. M. Murburg (Washington, DC: American Psychiatric Press, 1994), 253–76.

30. B. A. Van der Kolk, J. C. Perry et al., "Childhood Origins of Self-Destructive Behavior," *American Journal of Psychiatry* 148, no. 12 (1991): 1665–71.

31. M. Seligman, *Helplessness: On Depression, Development and Death* (New York: W. H. Freeman, 1992).

32. B. A. Van der Kolk, "The Compulsion to Repeat the Trauma."

33. J. D. Bremner, "Does Stress Damage the Brain?" *Biological Psychiatry* 45, no. 7 (1999): 797–805; J. D. Bremner and M. Narayan, "The Effects of Stress on Memory and the Hippocampus throughout the Life Cycle: Implications for Childhood Development and Aging," *Development and Psychopathology* 10, no. 4 (1998): 871–85; L. Cahill, *The Neurobiology of Emotionally Influenced Memory. Implications for Understanding Traumatic Memory in Psychobiology of Posttraumatic Stress Disorder*, ed. R. Yehuda and A. C. McFarlane (New York: Academy of Sciences, 1997), 821:238–46; and M. B. Stein, C. Koverola et al., "Hippocampal Volume in Women Victimized by Childhood Sexual Abuse," *Psychological Medicine* 27, no. 4 (1997): 951–59.

34. B. A. Van der Kolk and M. Greenberg, "The Psychobiology of the Trauma Response: Hyperarousal, Constriction, and Addiction to Traumatic Reexposure," in *Psychological Trauma*, ed. B. A. Van der Kolk (Washington, DC: American Psychiatric Press, 1987), 63–88; and B. A. Van der Kolk, M. Greenberg et al., "Endogenous Opioids, Stress Induced Analgesia, and Posttraumatic Stress Disorder," *Psychopharmacology Bulletin* 25 (1989): 417–42.

35. J. L. Herman, *Trauma and Recovery.*
36. B. James, *Handbook for Treatment of Attachment Trauma Problems in Children* (New York: Lexington Books, 1994).
37. S. L. Bloom, "Beyond the Beveled Mirror: Mourning and Recovery from Childhood Maltreatment," in *Loss of the Assumptive World: A Theory of Traumatic Loss*, ed. J. Kauffman (New York: Brunner-Routledge, 2002).
38. R. Janoff-Bulman, *Shattered Assumptions: Towards a New Psychology of Trauma* (New York: Free Press, 1992).
39. T. Pyszczynski, S. Solomon et al., *In the Wake of 9/11.*
40. L. Janis, "Decision Making under Stress," in *Handbook of Stress: Theoretical and Clinical Aspects*, ed. L. Goldberger and S. Breznitz (New York: Free Press, 1982), 69–87.
41. T. Pyszczynski, S. Solomon et al., *In the Wake of 9/11.*
42. I. E. P. Menzies, "A Case Study in the Functioning of Social Systems as a Defense Against Anxiety," in *Group Relations Reader I*, ed. A. D. Colman and W. H. Bexton (Washington, D.C.: A. K. Rice Institute Series, 1975).
43. B. A. Van der Kolk, P. Brown et al., "Pierre Janet on Post-traumatic Stress," *Journal of Traumatic Stress* 2 (1989): 365–78; and B. A. Van der Kolk and O. van der Hart, "Pierre Janet and the Breakdown of Adaptation in Psychological Trauma," *American Journal of Psychiatry* 146 (1989): 1530–40.
44. J. Goldstein, *The Unshackled Organization* (Portland, OR: Productivity Press, 1994); F. Masterpasqua and P. A. Perna, eds., *The Psychological Meaning of Chaos: Translating Theory into Practice* (Washington, D.C.: American Psychological Association, 1997); and B. A. McClure, *Putting a New Spin on Groups: The Science of Chaos* (Mahway, NJ: Erlbaum, 1998).
45. S. L. Bloom, *Creating Sanctuary: Toward the Evolution of Sane Societies* (New York: Routledge, 1997).
46. R. Abramovitz and S. L. Bloom, "Creating Sanctuary in a Residential Treatment Setting for Troubled Children and Adolescents," *Psychiatric Quarterly* 74, no. 2 (2003): 119–35; S. L. Bloom, "Creating Sanctuary: Healing from Systematic Abuses of Power," *Therapeutic Communities: The International Journal for Therapeutic and Supportive Organizations* 21, no. 2 (2000): 67–91; and S. L. Bloom, J. F. Foderaro, and R. A. Ryan, *S.E.L.F.: A Trauma-Informed, Psychoeducational Group Curriculum* at www.sanctuaryweb.com, 2006.
47. S. L. Bloom, "By the Crowd They Have Been Broken, by the Crowd They Shall Be Healed: The Social Transformation of Trauma," in *Post-Traumatic Growth: Theory and Research on Change in the Aftermath of Crises*, ed. R. Tedeschi, C. Park, and L. Calhoun (Mahwah, NJ: Erlbaum, 1998).

48. S. L. Bloom, "Every Time History Repeats Itself the Price Goes Up: The Social Reenactment of Trauma," *Sexual Addiction and Compulsivity* 3, no. 3 (1996): 161–94; and S. Bloom and M. Reichert, *Bearing Witness: Violence and Collective Responsibility* (Binghamton, NY: Haworth Press 1998).

49. D. Corn, *The Lies of George W. Bush: Mastering the Politics of Deception* (New York: Crown, 2003); J. W. Dean, *Worse Than Watergate: The Secret Presidency of George W. Bush* (New York: Little, Brown, 2004); and J. W. Dean, *Conservatives without Conscience* (New York: Viking, 2006).

50. R. A. Clarke, *Against All Enemies: Inside America's War on Terror* (New York: Free Press, 2004); R. Susskind, *The Price of Loyalty: George W. Bush, the White House, and the Education of Paul O'Neill* (New York: Simon and Schuster, 2004).

51. E. Alterman, *What Liberal Media? The Truth about Bias and the News* (New York: Basic Books, 2003); C. Brown, ed., *Lost Liberties: Ashcroft and the Assault on Personal Freedom* (New York: New Press, 2003); and N. Chang, *Silencing Political Dissent: How Post-September 11 Anti-Terrorism Measures Threaten Our Civil Liberties* (New York: Seven Stories Press, 2002).

52. J. Conason, *Big Lies: The Right-Wing Propaganda Machine and How It Distorts the Truth* (New York: St. Martin's Press, 2003); W. D. Hartung, *How Much Are You Making on the War, Daddy?: A Quick and Dirty Guide to War Profiteering in the Bush Administration* (New York: Nation Books, 2003); and R. C. Leone and G. Anrig, eds., *The War on Our Freedoms: Civil Liberties in an Age of Terrorism* (New York: Public Affairs, Perseus, 2003).

53. C. Johnson, *The Sorrows of Empire: Militarism, Secrecy and the End of the Republic* (New York: Metropolitan Books, 2003); N. Mailer, *Why Are We at War?* (New York: Random House, 2003); and S. Rampton and J. Stauber, *Weapons of Mass Deception: The Uses of Propaganda in Bush's War on Iraq* (New York: Tarcher, 2003).

54. J. Surowiecki, *The Wisdom of Crowds: Why the Many Are Smarter Than the Few and How Collective Wisdom Shapes Business, Economics, Societies and Nations* (New York: Doubleday, 2004).

55. F. Braudel, *A History of Civilizations* (New York: Allen Lane, 1994), 29.

Part 4

Trauma and Recent Cultural History

9

The Snake That Bites

The Albanian Experience of Collective Trauma as Reflected in an Evolving Landscape

Michael L. Galaty, Sharon R. Stocker, and Charles Watkinson

We first visited the Balkan nation of Albania in 1994–95, a few short years after the fall of Communism, and have worked there, conducting archaeological research, since 1998. We have had the fascinating privilege of watching a nation adjust, sometimes painfully, to very new and different realities. Gone are the days of totalitarian dictatorship, and the suffering that came with it. Gone, too, are the stability and security, largely illusory, born of a command and control economy and xenophobic foreign policy.

Albanians have good reason to desire stability and security, and they have experienced their fair share of suffering. Throughout their history they have been overrun countless times by external powers—Romans, Turks, Nazis, among others. In seeking protection from such outside forces, Albanians threw their support behind Enver Hoxha, who ruled the country following World War II until his death in 1985.

> Through all the centuries of their history, the Albanian people have always striven and fought to be united in the face of any invasion which threatened their freedom and the motherland. This tradition was handed down from generation to generation as a great lesson and legacy, and precisely herein must be sought one of the sources

> of the vitality of our people, of their ability to withstand the most ferocious and powerful enemies and occupiers and to avoid assimilation by them.[1]

When we were asked to contribute to this volume, our first reaction was to ask "Why?" "How might a paper written by archaeologists who work in Albania fit into a book about trauma?" Having heard one of us describe Albanian history, the editors thought we might have something interesting to contribute to the conversation, and perhaps they were right. Rarely, it seems, do we consider trauma's impact on entire groups of people. More rarely still can we trace the impact of repeated traumas through time. We will argue in this chapter that the Albanian experience of trauma—of repeated invasions, cultural and political domination, and collective suffering over the course of several thousand years—led to their rather peculiar behavior in the latter half of the twentieth century. As archaeologists, we build our argument with reference to the landscape and material culture of Albania, which embodies and communicates in symbolic form that nation's deeply traumatic history.

Albania, located along the western flank of the Balkan peninsula, between the former Yugoslavia and Greece, is a small, mountainous country (see figure 9.1). Its history and archaeology are not well-known to outsiders, including most Mediterranean specialists. Foreign archaeological explorations were conducted in the early part of the twentieth century, first by Austrian scholars and later by Italians and the French. During this period, a fledgling, indigenous Albanian tradition of scientific archaeological investigation appeared, though its growth and maturation only occurred following World War II, when the Communists, led by Enver Hoxha, gained control of the country.

Hoxha considered the past an important resource and encouraged the study of it.[2] History and archaeology were ideological tools wielded by his regime to support Marxist dogma, nationalist agendas, and a totalitarian system of government. During these years, Albania's leadership cultivated political and economic isolation and the country's citizens were allowed only intermittent contact with the outside world. In fact, in Hoxha's Albania the sociopolitical system, which subsumed and sponsored all scientific ventures (including archaeology), evolved in a near vacuum, one sealed and maintained by the Communists. In so tightly controlled an environment, Hoxha and his party managed to turn all forms of material culture to their purposes—including, as we will argue, whole landscapes.

The impact of human behaviors on regional landscapes and vice versa is a topic of intense discussion in the social sciences generally. Study of a specific type of monument—bunkers—and a particular artifact—a second century AD architectural block—will allow us to underscore several of the many problems faced by landscape archaeologists. In addition, we will demonstrate

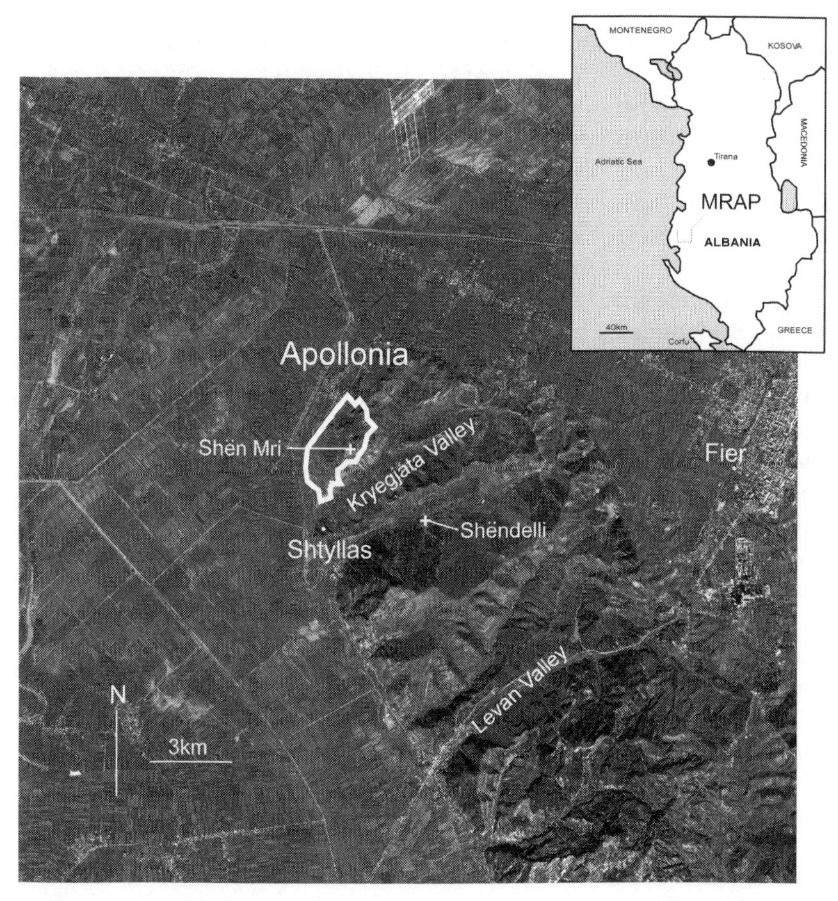

FIGURE 9.1. Map of the MRAP study region with sites mentioned in the text (Galaty).

that trauma is one among many human emotional experiences, including domination and resistance, that might conceivably be written into the fabric of an archaeological landscape.

Our research in Albania makes apparent the mutable and dynamic character of regional landscapes, wherein artifacts and whole monuments move about or are destroyed (often to be reassembled or reused at a later date), lose identity, and gain new significance. Consequently, given that landscapes are indeed constantly changing, the many possible former symbolic meanings of ancient artifacts and monuments, those still present in modern landscapes, are not, as many archaeologists readily admit, easily established. As our examples will make clear, people's understanding of what particular monuments mean can be transformed within the space of a few years, even after decades of symbolic stability.

We would further submit that archaeologists too often consider built landscapes and associated monuments from the top down, as testaments to elite power and domination; rarely do they search for signs or symbols of resistance. As Paynter and McGuire have written, "Elites commonly express dominant ideologies in a material culture that is grand and lasting, and more likely to be found by archaeologists. We should be hesitant, however, to assume that [all people] readily accepted such dominant ideologies."[3] Both bunkers and the architectural block, as we will describe more fully in a moment, have at key points in their history served as symbols of domination and control, and we would argue trauma, but have become or are becoming icons of resistance to and contempt for the toppled Communist government, that is, healing. As a result, strategies of both domination and resistance should, wherever possible, be incorporated into models of regional landscape change, for it is the interplay of such strategies that appears to create, in large part, the physical and cultural landscapes that are of interest to archaeologists, and social scientists generally.

In studying an Albanian regional landscape at multiple scales of analysis and from different perspectives, we are able to access much, to again quote Paynter and McGuire (13), of the "full theatre" of power and its workings. The examples we describe further will allow us to consider how the Communist dictatorship, over the course of nearly fifty years, managed to manipulate the material record to help impose its ideological vision on the Albanian population. We will also discuss the consequences of these strategies of domination, by describing what has become of Communist material culture in the post-Communist period, almost twenty years after the overthrow of hard-line regimes throughout Eastern Europe. In this chapter, we focus on a smaller geographic area—not the whole country per se, but two contiguous valleys—and address both short-term and long-term change. By reducing the scale of analysis and widening the time span, it becomes possible to expose more specific and fine-grained examples of popular reaction to what may be construed as a dominating material discourse—and perhaps even to suggest reasons why the Albanian dictatorship's strategy of domination eventually faltered.

Bunkers

The valleys of interest here, Kryegjata and Shtyllas, lie in central Albania, in an upland region known as the Mallakastra, overlooking the Adriatic (see figure 9.1). From the sixth century BC to the mid-first millennium AD, the classical city of Apollonia, an *apoikia* or "daughter colony" founded by Corcyran Greeks with assistance from their mother city, Corinth, dominated this landscape. The most prominent feature of the archaeological site of Apollonia is now a large second century AD civic building (a *bouleuterion,* often referred to as the "Monument of Agonothetes," after the name of its donor), partially reconstructed during the Communist period (see figure 9.2). In fact, one positive consequence of the Hoxha government's policies was the privileging of archaeological research, even though the results were often employed in Communist party propaganda (Galaty and Watkinson 8–12). For example, under Hoxha, funds were made available for the excavation and restoration of ancient monuments, funds not available prior to "liberation" and certainly not available today, given the present economic climate. The bouleuterion at Apollonia was, therefore, "restored" (as a façade, for purposes of display) and used to reinforce the party line of "giving the land and their history back to the people." In medieval times, a church and monastery were erected nearby, reusing classical building material, and most recently an archaeological dig-house has also been built, in the architectural style of the monastery.

> [But] in the land that had previously belonged to the Illyrians and had now been handed down to their descendants under the name of Albania . . . the old pyramid [of Cheops] spawned not thousands,

FIGURE 9.2. **Bouleuterion (or "Monument of Agonothetes") (Galaty).**

> but hundreds of thousands of little ones. They were called bunkers, and each of them, however tiny it may have been in comparison, transmitted all the terror that the mother of all pyramids had inspired, and all the madness too.[4]

Mushrooming from every hillside in the Mallakastra survey area, as in Albania as a whole, concrete bunkers, gun emplacements, and pillboxes (collectively referred to as *bunkeri* in Albanian) are a main built legacy of Communism in the rural landscape (see figure 9.3). More so even than monumental sculptures, bunkers are the material symbols of the Albanian dictatorship. Although initially allied with the USSR and later with China, in the late 1970s Albania broke with all its Communist allies, inaugurating a period of paranoid xenophobia and economic stagnation. The bunkers were a key feature of this isolationist policy, designed to provide a last-ditch defense against foreign invasion. Beginning as early as 1967 and continuing through 1986, an estimated four hundred thousand to eight hundred thousand concrete bunkers—each reinforced with thirteen layers of steel—were constructed at the personal command of Enver Hoxha and with the approval of the Albanian Politburo. This amounts, on average, to one bunker for every four to five Albanians (at population levels of 1989). Made wary by the Soviet invasion of Czechoslovakia and suspicious of the new military junta in Greece, Hoxha believed that Albania might someday have to fight a

FIGURE 9.3. Pillbox bunker (Galaty).

two-front war and insisted that the country prepare for the simultaneous invasion of as many as eleven airborne divisions. During this period, prefabricated bunker construction alone accounted for an estimated 2 percent of net material product.

In theory, bunkers were the key to Hoxha's defensive military strategy, one designed to deflect looming, though probably imaginary, external threats. The history of Albania is one of foreign occupation—by ancient Greeks, Romans, Bulgarians, Venetians, Ottomans, Italians, Austrians, Germans, Serbs, and modern Greeks. Hoxha took advantage of the determination of the Albanian people to resist further assimilation or subjugation to foreign powers, and bunkers emerged as a material expression of that resistance. In practice, however, their symbolic role as instruments of internal domination was perhaps more significant. On a journey through the Albanian landscape, bunkers seem to be everywhere: they appear around every corner, in all parts of the country. Darrow, a journalist, noted that "it doesn't take a psychologist to perceive the impact of such imagery."[5] Indeed, Paynter and McGuire (9) have argued, following other authors, that "patterned large-scale construction has a disciplinary potential as a means of familiarizing a population with a given order of rule." Following a similar line of reasoning, the Western popular press typically emphasizes the seemingly oppressive nature of bunkers. Reporting for CNN, Darrow describes a "blight not only on the landscape but on the collective memory and perceptions of the people" and contends that bunkers are symbols of Communism's attempt to build a "bulwark against the invasion of ideas and influences from the outside world."[6] Bunkers, it seems, were and still are manifestations of Hoxha's philosophy of leadership: "If we slackened our vigilance even for a moment or toned down our struggle against enemies in the least, they would strike immediately like the snake that bites you and injects its poison before you are aware of it."[7]

Discussing the megalithic burial mounds, stone circles, and linear earthworks of prehistoric northwest Europe, various archaeologists have argued that these monuments, by structuring daily movements through the landscape, played an important role in normalizing and legitimizing unequal power relations. In Communist Albania, it can be argued, bunkers were meant to play much the same role. They created what might be described as a "siege mentality," causing many Albanians under Hoxha to believe (or, at least, claim that they believed) in the possibility of foreign invasion, and therefore also in the need for a common, though expensive, defense. In addition, beyond the initial construction itself, the regular maintenance of bunkers was organized on a family-by-family basis by the local party secretary or "brigadier," thereby binding local communities into an infrastructure of authority extending all the way to the center, in Tirana.

If, as material symbols, bunkers were meant to unite the country in common cause against a predatory outside world, in retrospect most Albanians describe them as blatant symbols of repression. Albanians today will often insist that they had always considered bunkers to be symbols of intimidation and control rather than of nationalist unity (though this is, after the fact, difficult to substantiate). For instance, in his novel *The Pyramid* (first published in 1992), Ismail Kadare, Albania's most esteemed living author, used bunkers to symbolize the brutality and control of the Hoxha regime.[8] Likewise, Albanian filmmaker Kutjim Çashku wrote and directed *Kolonel Bunker*, a movie, released in 1996 and set in the 1970s, that depicts the absurdity of Hoxha's "bunkerization" program. Asked about bunkers, Çashku remarked, "I feel the bunkers has [*sic*] been a symbol of totalitarianism for myself, because it was first of all the isolation psychology."[9] Today, bunkers appear to have lost whatever previous meaning or meanings they were meant to have by the Hoxha government, but, in so doing, they have assumed new meanings. They may be ignored or, worse (better?) yet, ridiculed. In fact, small carved soapstone "pillbox" bunkers, which function as pencil holders or ashtrays, have become Albania's leading tourist souvenir, an irony not lost on the Albanians themselves. In an admittedly unsystematic and unscientific survey, we posted to an Albanian Internet site the question "What do bunkers mean to you?" Albanians who responded, primarily young adults in their late twenties and early thirties, used such terms as "disharmony," "useless," "strange," and "wasteful"—as well as mentioning that bunkers were the place where teenagers went to behave "promiscuously." In a country with no cars, the bunker assumed the role of the proverbial "backseat!"

In addition to building bunkers, the Hoxha regime undertook other massive campaigns designed to "improve" the Albanian landscape. Dean Rugg, in a critique of socialist landscape theory, describes several of these: the canalization of the swampy Myzeqe Plain, the complete collectivization of agriculture, the damming of the Drin River, the construction of forty-one new, planned urban centers, and the reconstruction of older cities, such as Tirana, according to a Marxist blueprint.[10] As with the manufacture and maintenance of bunkers, these projects also served to reinforce existing power structures and to this day still remind Albanians—especially those sent to forced-labor/reeducation camps—of the control once exercised over their lives by the Communist government. However, perhaps because they represent economic (that is, infrastructural) development, these types of large-scale projects do not appear to have held for Albanians nearly the same symbolic force as did Hoxha's more jolting and offensive manipulations of regional landscapes. For example, sculpted in giant letters on a mountain slope facing the city of Berat, and still visible today, was the dictator's first name—ENVER. The word loomed over the countryside and the message was simple and unavoidable: Resistance is futile.

Since the overthrow of the hard-line Communist regime in Albania, monumental sculptures have been defaced, torn down, and destroyed. For example, the giant statue of Hoxha that once stood in Skanderbeg Square in downtown Tirana was toppled in 1991. The square is now dominated by bumper cars. In the port city of Durrës, Communist-era statues were dismantled and removed to the local archaeological museum, where they now sprawl behind the building, rusted and forgotten. The solidity of the concrete bunkers, especially the very large ones, has made them somewhat more intractable. In cities, where land is at a premium and heavy lifting gear is available, bunkers have been physically uprooted, but in rural Mallakastra most bunkers have simply been abandoned. They are becoming absorbed into the landscape. Some are being reused for practical purposes, such as sheepfolds, dog kennels, or kilns, and in cities such as Durrës there have been some more adventurous experiments—for instance, one large bunker, located on the beach, has been turned into the Restaurant Bunkeri. But for the most part bunkers now lie derelict. As Knapp and Ashmore, as well as Barrett, argue, "Less catastrophic fates [than obliteration] for monuments and landscapes may be no less socially profound, inasmuch as what we might call benign neglect may signal a fundamental change in the social perception of the landscape, its past, and the society it represents."[11]

Beyond Bunkers

In addition to building material symbols of its dominance, Hoxha's regime sought to destroy the representations of other, potentially dissident ideologies. Taking their cue from the Chinese Cultural Revolution, during the 1960s and into the 1970s, local Communist parties razed or otherwise destroyed an estimated 95 percent of the 2,169 churches and mosques in Albania. The church of Shëndelli (St. Ilias) in the Shtyllas Valley was one victim of these purges. In 1998, all that remained of the church was a low mound of rubble, with only the foundations still visible.

One large masonry block remained above after the devastation, apparently retrieved from the wreckage of the church (see figures 9.4 and 9.5). During our fieldwork we were able to link this block stylistically to the Roman-period bouleuterion at Apollonia. Several similar blocks, located in the museum courtyard at Apollonia, bear identical carved motifs, as do ancient blocks built into the monument, as reconstructed during the Communist period. Despite its origin at a site in another valley, and at least a five-mile journey by road, this block had been incorporated into the church, possibly at the time of its construction sometime prior to the last quarter of the nineteenth century.[12] In fact, such reuse of ancient material is not uncommon in this region. The Byzantine monastery and church at Shën Mri is constructed in large part from the remains, both building stones and ceramic tiles, of the nearby ruins of Apollonia. A large nineteenth-century farmhouse in the Kryegjata Valley incorporates massive cut-stone blocks,

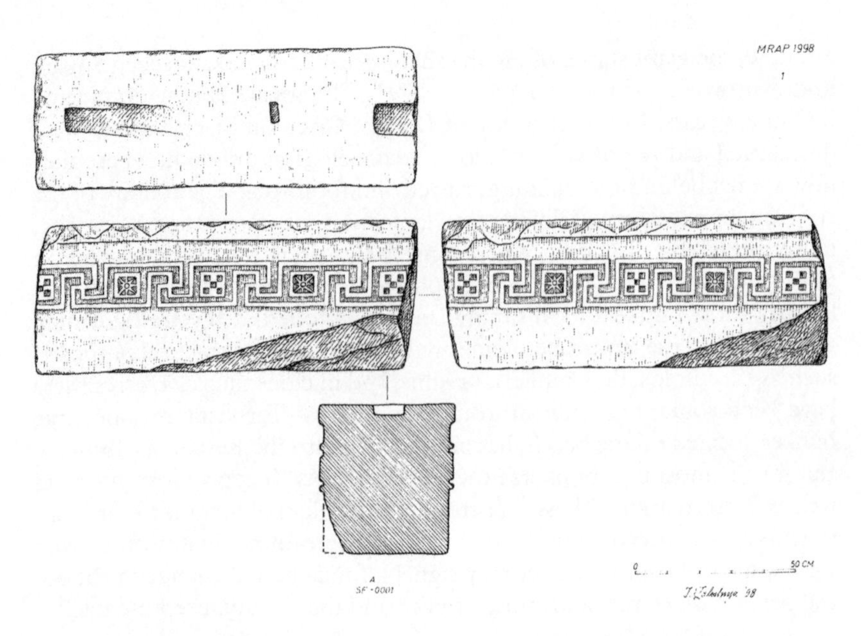

FIGURE 9.4. Architectural block at Shëndelli (drawing by Ilir Zaloshnija).

presumably from Apollonia. Most dramatically, a lone standing column on a nearby hill at Shtyllas ("column" in Albanian) is all that remains of a classical temple, mentioned by Cyriacus Anconitanus already in the fifteenth century and later described by numerous nineteenth-century English visitors.[13] Leake also described how masonry had been taken almost fifty miles away from Apollonia by cart to be built into the new citadel of the Ottoman Pasha of Berat, Ibrahim Pasha, father-in-law to Ali Pasha's two eldest sons.[14]

> [I]nquiring of the monks, I was informed that no less than seventy cart loads had been taken from thence to build the new serái at Berát. Similar spoliations have been committed at the western temple [Shtyllas], and so recently that the excavation made to carry away the foundations, of which not a single stone is left, affords a very tolerable measurement of the length and breadth of the building. . . . One column standing in solitary grandeur is the only part of it that has been spared the Pasha's masons.[15]

At the site of Shëndelli, the ancient block, later part of a Christian church, has now found yet another function and meaning. On visits to the site in 1998 we often found flowers spread over and around it, and saw marks of burning. Villagers, both Christians and Muslims, come from Shtyllas, about a mile down the valley, to worship at the block, lighting candles and bringing fresh flowers. Around the site were marks of other religious practices: a safe (marked with the date 1991) had been brought for votive offerings, pits held deposits of clothing and personal items, and two tree stumps were covered with fragments of roof tiles; but at the center of all this activity was the block itself. Villagers claim that the feast day of Shëndelli had been regularly observed at the site, even under the Hoxha regime—despite the fact that, following Hoxha's antireligious campaign, which began in 1967, such activity was punishable by a long prison term and even death, if discovered. We have also learned that many of the inhabitants of the village of Shtyllas were related to, or were themselves, political dissidents, resettled by Enver Hoxha in this relatively remote rural area, far from urban centers of power. Internal exile was a common practice of the Hoxha government, designed to exercise social control and discourage illegal activities. For such Albanians, already considered enemies of the state, the act of worshipping at the block on the grounds of a destroyed church constituted an impressive act of resistance.[16]

Danny Miller and his colleagues have written that "the past is not a static, archaic residue, rather it is an inherited artifact which has an active influence in the present through the interplay of popular and officially

FIGURE 9.5. Architectural block at Shëndelli, in use as an altar (Galaty).

inscribed meanings."[17] In the block, we confront an artifact that has held various meanings, in various locations, over the course of its long existence. We would suggest that the block, unlike a bunker, has become one key element in an invented local tradition, one that joins the past to the present and, in the context of resistance, refers back to local, rather than central, systems of authority. It was originally carved for use as an architectural element in a large and imposing civic building, eventually making its way to Shtyllas to be built into the Shëndelli church. By incorporating into their church a block from a prominent, local site of evident antiquity, the builders of Shëndelli appropriated (or, in Miller and his colleague's terms "inherited") a powerful symbolic force, one that the Pasha of Berat is likely to have understood. Indeed, we can be certain that the father-in-law of his daughters, Ali Pasha of Ioannina, comprehended the political capital to be gained by associating himself with the Hellenic past.

Whatever their actual knowledge of the history of Apollonia, by using architectural elements from the site, the local religious leaders who originally built the church may well have recognized that the ancient site was an important structuring feature of the local landscape, a landmark. The modern worshippers at the site whom we interviewed were not aware of the direct link between this block and Apollonia, although they were conscious of its former use in the now-destroyed church. Certainly, both the builders of the church and the Pasha of Berat also had practical motives in reusing building materials from Apollonia: a cut, dressed stone of striking appearance to embellish their new church; and, in the case of the Pasha, prefabricated bricks and tiles, free for the taking. We do in fact possess documentary evidence indicating that Ali Pasha of Tepelena, the infamous "Lion of Ioannina," knew full well that by mining ancient sites he tapped into his country's glorious past and, at the same time, augmented his own reputation.[18] Ibrahim Pasha could hardly have been unaware of such attitudes, perhaps also communicated to the builders of Shëndelli. More than a century later, and following the destruction of the church, the villagers of Shtyllas, many of whom had already been exiled for their dissident views, again appropriated the past for their own symbolic uses. However, in picking the block, they had successfully, though perhaps unknowingly, connected their new traditions of resistance—through the practice of a particular religious ritual—to regional histories extending back almost two millennia. The irony is that a block that originated as part of a Greco-Roman monument, one "reconstructed" by Hoxha for the purpose of domination, was used to symbolize resistance at the site of a monument, a church, Hoxha himself had destroyed! Knapp and Ashmore have noted that "modernization of landscapes often leads to truncation and impoverishment of their living embodiment of memory, to a rupture in their 'cultural biography'" (10). If, therefore, we regard Communism as a truncation, what we now witness is a resumption of the block's cultural biography.

Discussion and Conclusion

The great river of life flows not evenly for all peoples. In places it crawls sluggishly through dull flats, and the monuments of a dim past moulder upon the banks that it has no force to overthrow; in others it dashes forward torrentially, carving new beds, sweeping away old landmarks; or it breaks into backwaters apart from the main stream, and sags to and fro, choked with the flotsam and jetsam of all the ages. Such backwaters of life exist in many corners of Europe—but most of all in the Near East. For folk in such lands time has almost stood still. The wanderer from the West stands awestruck amongst them.[19]

The English traveler, Edith Durham, who visited Albania at the beginning of the twentieth century, was struck by the lack of change. In this chapter, writing at the start of a new millennium, we have focused on a period of extremely rapid change. Whether it comes in long, slow cycles or as bursts, the nature and consequences of change are of central concern to archaeologists.

The Albanian landscape is an ever-changing entity, an immense artifact that has been itself constructed and reconstructed numerous times throughout its history and prehistory. Both Albanian individuals and Albanian (and foreign) governments—the Hoxha government included—have participated actively in ongoing processes of regional landscape change. Sometimes the pace of this change has been gradual and accretionary, as in the slow, stealthy looting of archaeological sites, and at other times abrupt and violent:

> When the [Hoxha] régime finally came to an end in 1991, there followed an extraordinary orgy of destruction and vandalism. As if the world had come to an end and there would be no future needs, vineyards and orchards were destroyed, cooperative buildings razed to the ground; school windows, furniture and books demolished; machinery broken, and the entire rural telephone system ripped out.[20]

Either way, with the end of Communism, the Albanian landscape—especially the built landscape—is being transformed, though this time we, as archaeologists, may bear witness to its transformation. What is most remarkable is just how rapidly evidence of the Communist period is being erased from the countryside (and, we might note, from the Albanian consciousness: a generation has now reached maturity that did not experience Albania under Hoxha). Obvious symbols of Communism, such as statues, have already nearly disappeared. Those monuments that have resisted destruction or removal, such as bunkers, are, as symbols, being deconstructed and reassembled, their new meanings designed to serve a very different, Capitalist (as opposed to Communist) Albania. In a personal communication (May 1999), Helen Regis, a social anthropologist who studies contemporary human reactions to modern landscapes, noted:

> The restaurateur who sells "bunkeri" burgers is also joining the past to the present, although in a different way. He exerts his own power over the production of history by juxtaposing the post-communist present to the impotent past of Hoxha's dictatorship. . . . If used as souvenirs, they [bunkers] have not ceased to communicate, but are being employed to communicate a very different message: a self-deprecating, post-communist kitsch aesthetic which recuperates the past as "heritage" through the idiom of mockery.

That ancient monuments, such as henges and stone circles, "structured ancient landscapes" or "legitimized power relations" may indeed be true; yet we will never know exactly what they meant to prehistoric Britons. Unlike strictly archaeological examples, however, we can demonstrate the shifting meaning of bunkers as symbols to Albanian people through time, from the period of their construction, during their use, and at their destruction. Whereas they once represented (or were meant to represent) social solidarity and defense, they now may symbolize for the Albanians past injustices, those of an overthrown political regime, one now openly mocked. The idea that bunkers embodied Communist ideology in material form and thus once lent structure and meaning to the landscape supports the possibility that large, visible monuments functioned similarly in prehistory. The greater difficulty lies in objectively reestablishing the particular symbolic meaning(s) that ancient monuments held at particular points in time. We would argue that this exceedingly difficult task may not be able to be accomplished in totally prehistoric contexts by simple reference to a monument's "prominent" location, nor by uncritical comparison with "similar" monuments from very different times and/or places. For example, as a colleague, Janet Rafferty, has pointed out to us (personal communication, May 1999), the bunkers and gun emplacements along the British coast, at Dover and elsewhere, possess for the British people a very different meaning than their nearly identical counterparts in Albania have for Albanians.

If the Albanian material record is any guide, establishing associations between monument and meaning is certainly possible, albeit very difficult; and the possibility of doing so depends to a great extent on an accurate understanding of the particular, idiosyncratic archaeological region and record under study. We would argue that landscape archaeologists all too often merely assume the symbolic meaning of prominent monuments (most often as symbols of dominance), without actively considering other, alternative, though equally valid, possibilities. Consequently, if the archaeological study of landscapes is to become more sophisticated, we must continue to devise methods for establishing the symbolic meaning(s) of ancient monuments.

Landscapes represent open and dynamic symbolic systems that are layered in meaning, a "palimpsest." The various components of any built landscape—those features (artifacts and monuments) that lend the system structure and cultural significance—may depart, be destroyed or disappear, or be completely reinterpreted; their visibility and importance may wax and wane. Bunkers were forcibly thrust into the Albanian landscape and the consciousness of the Albanian population, and are now just as rapidly exiting (or, in some cases, are being functionally and/or symbolically reborn). They will indeed be preserved to some extent in the archaeological record and, as we have argued, we can document what their meaning is for contemporary Albanians, but we cannot readily predict how they will be perceived in the future. The block has already shifted location and meaning a minimum of three times in a use-life spanning some eighteen centuries. It encapsulates and encodes the history of Shtyllas, Mallakastra, and in fact Albania itself, representing glory for some (ancient Greeks and Romans), repression for others (the local Illyrians, perhaps), then defiance, and now a new beginning. Considered together, bunkers and the block—though functioning in the present at very different spatial scales—serve to illustrate that it is the interaction, through time, of many different actors, agents of both control and resistance, that transforms a regional landscape and with it the regional archaeological record. In Albania, it appears to be the interplay of power relations that has driven and is driving evolutionary change—and this in what was supposedly the most oppressive of modern dictatorships. As a result, it appears quite probable that study of domination alone will not allow adequate explanation of regional landscape change.

We have sketched two examples of the ways in which material culture of the past, both recent and distant, has been, and is being, used and reused in the valleys of central Albania in the immediate environs of the ancient site of Apollonia. We have argued that the bunkers and the block are material representations of two competing ideologies—domination and resistance, the cause of and a response to trauma. Viewed over the long term, and given a regional archaeological perspective, it certainly seems that particular monuments—the ruins of Apollonia, the block and the site of the destroyed church—are more enduring symbols than the more numerous bunkers. It is ironic that, at a local level, Enver Hoxha's material strategy for domination should have eventually failed, because the monuments—bunkers—that came to symbolize his government and its policies lacked a history. Archaeological sites such as Apollonia were restored and opened to the public under his regime, and substantial amounts of money were invested in heritage management and museum construction; Hoxha cared about the past and actively cultivated archaeological research. Bunkers have become Hoxha's material legacy, and because bunkers have no past—no symbolic capital on which contemporary Albanians may draw—that legacy, as it is inscribed on the landscape, may not be a lasting one. Bunkers are allowed to recede into the landscape, may occasionally be reused for practical purposes (as in the case of the Restaurant Bunkeri), but for Albanians now assume meanings other than the intended. On the other hand, it might also be suggested that bunkers continue, in Western eyes at least, to represent a striking success of Hoxha's policy, a resonant symbol in the Western world of the alterity of Albania and its unique position in a modern Europe.

In conclusion, we have tried to demonstrate that the Albanian historical consciousness, one marked by an inordinate degree of individual and national trauma, was a direct cause of the seemingly bizarre behavior of the repressive Hoxha government. In the end, Hoxha did not relieve his people's sense of suffering, rather he added to it. Bunkers are a material symbol of the traumatic Hoxha years; the Shëndelli block a glorious reminder of the power of human beings to resist oppression and to heal.

Notes

This chapter is a slightly changed version of a paper originally published as "Beyond Bunkers: Dominance, Resistance and Change in an Albanian Regional Landscape" in the *Journal of Mediterranean Archaeology* 12, no. 2 (1999): 197–214.

1. Enver Hoxha, *Laying the Foundations of the New Albania: Memoirs and Historical Notes* (Tirana: Nentori Publishing House, 1984), 11.

2. See the discussion in Michael L. Galaty and Charles Watkinson, "The Practice of Archaeology under Dictatorship," in *Archaeology under Dictatorship*, ed. Michael L. Galaty and Charles Watkinson (New York: Kluwer Academic/Plenum Press, 2004), 1–18. Hereafter page number(s) cited in the text.

3. Robert Paynter and Randall H. McGuire, "The Archaeology of Inequality: Material Culture, Domination, and Resistance," in *The Archaeology of Inequality*, ed. Randall H. McGuire and Robert Paynter (Oxford: Blackwell, 1991), 10. Hereafter page number(s) cited in the text.

4. Ismail Kadare, *The Pyramid*, trans. David Bellos from the 1992 French version of Jusuf Vrioni; Albanian title: *Pluhuri Mbretëror* (New York: Vintage, 1996).

5. Siobhan Darrow, "Albania Littered with Symbolism—In the Form of Bunkers," *CNN Interactive*, March 18, 1997. <http://207.25.71.90/WORLD/9703/18/albani.bunkers/> (accessed November 21, 2007).

6. Compare similar reporting by Sylvia Pogiolli, "Albania—Where Democracy Building Went Bust," in *Nieman Reports* 51, no. 2 (1997):

46–48; and Carol J. Williams, "Bunker Mentality Survives in Albania," *Ann Arbor News*, November 24, 1994.

7. Enver Hoxha, *Selected Works, 1966–1975*, vol. 4 (Tirana: Nëntori Publishing House, 1982), 562, as quoted in James S. O'Donnell, *A Coming of Age: Albania under Enver Hoxha*, East European Monographs, Boulder, no. 517 (New York: Columbia University Press, 1999), 218.

8. See note 4.

9. Quoted in Darrow, "Albania Littered with Symbolism."

10. Dean S. Rugg, "Communist Legacies in the Albanian Landscape," *Geographical Review* 84, no. 1 (1994): 59–73.

11. A. Bernard Knapp and Wendy Ashmore, "Archaeological Landscapes: Constructed, Conceptualized, Ideational," in *Archaeologies of Landscape: Contemporary Perspectives*, ed. Wendy Ashmore and A. Bernard Knapp (Oxford: Blackwell, 1999), 19, hereafter page number(s) cited in the text; and John C. Barrett, "The Mythical Landscapes of the British Iron Age," in *Archaeologies of Landscape: Contemporary Perspectives*, ed. Wendy Ashmore and A. Bernard Knapp (Oxford: Blackwell, 1999), 253–265.

12. The earliest historical reference to this church is by the traveler Alfred Gilliéron, *Grèce et Turquie: Notes de Voyage* (Paris: Librairie Sandoz et Fischbacher, 1877).

13. Cyriacus Anconitanus, *Inscriptiones, seu Epigrammata Graeca, et Latina reperta per Illyricum a Cyriaco Anconitano apud Liburniam Designatis locis, ubi quaeque inventa sunt cum Descriptione Itineris* (Rome, 1747).

14. Pasha is the honorary title awarded to military and civilian officials by the Ottoman sultan; pashas typically governed a provincial territory called a "pashalik."

15. That's John Martin Leake, *Travels in Northern Greece, in Four Volumes* (London: J. Rodwell, 1835), 373.

16. In 2005 the rebuilding of the church was completed. The block now sits on the ground near one of the church's new walls.

17. Daniel Miller, Michael Rowlands, and Christopher Tilley, eds., *Domination and Resistance* (London: Routledge, 1989), 4.

18. See, for example, Peter O. Brøndsted, *Interviews with Ali Pacha of Joanina in the Autumn of 1812; with some Particulars of Epirus, and the Albanians of the Present Day*, ed. J. Isager (Athens: Danish Institute at Athens, 1999).

19. Edith Durham, *High Albania: A Victorian Traveller's Balkan Odyssey* (originally published 1909, reissued by Sterling Publishing, New York, 2001).

20. Clarissa de Waal, "Decollectivisation and Total Scarcity in High Albania," in *After Socialism: Land Reform and Rural Social Change in Eastern Europe*, ed. Ray Abrahams (Oxford: Berghahn Books, 1996), 173.

10

Traumatic Life

Violence, Pain, and Responsiveness in Heidegger

Eric Sean Nelson

Introduction

The issue of trauma, both its causes and appropriate ethical responses to it, is already implicit in discourses on violence and pain. Approaching the question of suffering, if often only to excuse or ignore it, discourses on violence are inevitably haunted by—even in their avoidance and defacement of—the possibility of an irreparable suffering. The unfolding of such an irredeemable trauma, of a wound without healing, can be traced in the works of the German philosopher Martin Heidegger (1889–1976), who remains today both a central figure of twentieth-century philosophy and inevitably controversial given his involvement with National Socialism in the early 1930s and his postwar silence.

Heidegger has been criticized for being oblivious to the other, including the suffering of the other, and ultimately the entire question of ethical alterity. Such criticism can be answered in part, as Jean-Luc Nancy has shown, insofar as Heidegger articulated the notion of self and other from the relational context or nexus (*Zusammenhang*) of the "with" (*mit*) and "the between" (*Zwischen*).[1] Heidegger's philosophy of the between, which includes his analysis of human sociality according to the existential category of "being-with," challenged traditional dichotomies of subject and object and disclosed the possibility of an originary ethos of dwelling that is open and responsive to things and world. Nevertheless, critical questions remain: Did Heidegger fail to notice the specificity of the ethical relation between self and other? Is his philosophy closed to the human origins of dialogical meaning? Does it foreclose the possibility that the self discovers itself obligated to respond to the other in his or her suffering and trauma?

Despite his unpardonable political commitments of the early 1930s, it might well be helpful to rethink such criticisms of Heidegger's thought in order to reconsider the question of trauma itself. Instead of constructing trauma according to frameworks, ideologies, and theodicies that explain, justify, and excuse it, thus losing sight of it as trauma, a careful reading of Heidegger reveals the possibility of responding to the difference of irretrievable suffering in its traumatic character. This chapter accordingly intends to determine how such responsiveness to pain, and even trauma, is intrinsic to Heidegger's thought.

Questioning Violence and Its Trauma

According to Heidegger, in his 1928–29 lecture-course *Introduction to Philosophy*, pain is intrinsic to all violence: "Alle Gewalt aber birgt in sich den Schmerz."[2] At first glance this quote might suggest the Stoic acceptance of and indifference toward violence and pain. It is perhaps only the cost of doing business in the world or of achieving greatness. Yet this passage may instead imply a recognition of the intrinsically painful and questionable character of violence, especially if interpreted in light of Heidegger's works of the mid- and late 1930s. These works, recently published, show Heidegger's increasing recognition of the problematic character of violence and power.

This approach might at first seem strange, given that Heidegger's language and thought have been repeatedly criticized for their violence. Nowhere does this violence seem more apparent than in the speeches and lecture-courses delivered during the period between 1933 and 1935. In his *Rectoral Address*, given after the Nazi seizure of power in 1933, Heidegger employed the language of *Being and Time* to embrace the "self-assertion of the German university" in support of the National Socialist subordination or forced synchronization (*Gleichschaltung*) of the university to its goals.[3] The scholarly interpretation of the *Rectoral Address*, which marks the height of Heidegger's support for National Socialism prior to his disillusionment and resignation from the leadership of the University of Freiburg, is not without its own ambiguities such that the significance and degree of Heidegger's complicity with National Socialism remains disputed. This is because Heidegger, based on his commitment to anticommunism and German traditionalism, initially supported and laid out his own vision of National Socialism and yet his thought became increasingly incompatible with central tendencies of actual National Socialist ideology and practice. Over the course of the 1930s, he challenges its biologism, gigantism, racism, worship of power, frenzied commitment to the total mobilization of society, and its "nihilistic" commitment to "value-thinking." Nevertheless, whatever we think about Heidegger, this crucial issue of Heidegger's politics can and should not be removed from thinking about Heidegger's thought and legacy, and it will rightfully always reemerge as a challenge to the ethical and political implications of his philosophy.

Even after Heidegger's resignation from the rectorship, his 1935 lecture-course *Introduction to Metaphysics* has been condemned as the pinnacle of violence in his thought, and not only because it is explicitly dedicated to the question of violence.[4] Instead of being a simple endorsement of violence, as the standard interpretation maintains, I argue that this lecture-course moves ambiguously between ethically minimizing the significance of violence and recognizing its pain. Heidegger's articulation of pain and loss in this work discloses an irreparable suffering that should be contrasted with other notions of trauma.

According to this argument, Heidegger does not stoically or heroically-tragically adopt the attitude toward violence and trauma that can be described as indifference toward much less a celebration and valorization of "traumatized life" in the manner of, for instance, the young Ernst Jünger or Carl Schmitt. In his early writings, Jünger explored and celebrated soldiers brutalized and traumatized by their experiences on the fronts of World War I and the possibility of their being agents of the transformation of society. Jünger described how their dislocated, mobile, and constantly threatened existence provided a model for a society of total mobilization of masses and resources, such that the Kantian distinction between the human being as an autonomous "end in itself" and a manipulated object or purely instrumental means would collapse. Schmitt, the leading jurist and political theorist of the Third Reich, ideologically justified National Socialism through the concept of the leader ruling directly and immediately through a "state of emergency" instead of through the complications, compromises, and weaknesses of liberal democracy. Schmitt advocated that the state create such a condition of emergency through the threats of internal and external enemies. Under such conditions of constructed crisis and enmity, power can be unreservedly asserted and threats conquered without the restraints of liberal institutions and values.

Jünger and Schmitt can be interpreted as envisioning life as inherently traumatized and necessarily for the sake of political mobilization. This trauma must be embraced with either a tragic-heroic affirmation or with the political realism that defines politics as the art of distinguishing and appropriately dealing with "friend" and "enemy." These reactions, however, use the shattered subject without addressing it. They are not responsive and answerable to the very phenomenon of trauma, and the shattering of the subject and its world that this violence involves. For trauma cannot be mastered by a "subject"; nor can it be removed through the negation of the subject. There is no "recovery" in the sense of a return to an original nontraumatized existence. Nor is there the emergence of a subject unchallenged by the trauma that is its past, even if the subject is reinterpreted as total mobility or as demoralized will to power.[5] Jünger and Schmitt accordingly do not recognize trauma precisely as trauma, that is, the other in her trauma as that which is "each time" her own in being shattered and becoming a question for herself. Heidegger's thinking of death in *Being and Time* and his thinking of violence in the *Introduction to Metaphysics* are connected in revealing a powerful alternative to the philosophy of the subject—and the nihilistic annihilation of the subject—for responding to the question of trauma.

For all Heidegger's possible complicity during this period, his thought—perhaps despite itself—challenges the ideological uses of traumatized life that he himself sometimes engaged in. Whereas violence, trauma, and pain are used to justify fascist ideology, and were celebrated in it, Heidegger's *Introduction to Metaphysics* reveals their aporetic, contradictory, and paradoxical character. The ambivalent movement of this work unfolds into an analysis of the ideological practices of National Socialism, such as its endorsement of value-thinking and its racist Social Darwinism as expressed in the misappropriation of the works of Friedrich Nietzsche by leading National Socialist intellectuals. It also gives rise to one of Heidegger's most controversial utterances—the hope of disclosing its philosophical "inner truth and greatness" (*IM* 152).[6]

Much has been written about the political background and ramifications of Heidegger's activities during these years. This painful context should be kept in mind as we turn to the question of this chapter: What does violence signify given Heidegger's turn against and "critical encounter" (as *Auseinandersetzung*, if not "critique" in the traditional sense) with violence, power, and domination beginning with the *Beiträge* of 1936 and continuing throughout his later thought? This chapter returns to the *Introduction to Metaphysics* as a crucial moment in transition in Heidegger's thinking in order to inquire into the pain and trauma inherent in its violence.

If this interpretive strategy is appropriate, then one may question whether Heidegger's thinking of violence and decision involves the decisionism and the violence of ontology that critics of Heidegger such as Levinas and Habermas propose.[7] Levinas criticized Heidegger's ontology for the constitutive role of violence in it. The structure of Dasein as a being who is concerned with its own existence reflects egotistic self-concern and the violent self-assertion of the will and *conatus*. Basic agonistic words such as Auseinandersetzung, *polemos*, and *Streit* echo the Darwinian struggle for existence. Levinas, as Robert Bernasconi has argued, consequently interprets Heidegger's thought as derivative of biologism and Social Darwinism.[8] Moreover, according to Levinas, being is a realm of conflict and war untouched by the transcendence of the ethical claim and thus ontology is inherently unresponsive to the pain, suffering, and trauma of the other.[9] Likewise, John Caputo has argued for the need to deconstruct the valorization of violence in Heidegger's *Introduction to Metaphysics* and, in another article, against the moral and phenomenological adequacy of Heidegger's understanding of pain.[10]

Despite the validity of Bernasconi's and Caputo's assessments, there is a need to reconsider the issue of violence and ontology in Heidegger's *Introduction to Metaphysics* and the works of the late 1930s. Heidegger's deepest confrontation with questions of violence occurs during this period, and his recognition of violence in this work calls for a more nuanced reading than the standard interpretations of what informs his critique of violence and power in the recently published writings of that era. In contrast to how his thinking of origins is portrayed by critics, rupture and trauma already occur in the origins that Heidegger's approach intends to reopen. Rather than submerging human existence in primordial origins and a totalizing and indifferent "Being," being (*Sein*) is interpreted through the formal indication of the human being as "Dasein" or "being there," that is, that being which is a question for itself as always already thrown, fallen, and dispersed in a world. Instead of celebrating the violence of being that the human being encounters, Heidegger begins to elicit possibilities for responding to this violence. In the face of the overwhelming and the uncanniness of existence, Dasein is in each case forced to respond in one way or another. This violence of being, history, and nihilation in general calls forth responsiveness to pain. That is, it can possibly bring forth a reply that would be constitutive of a different understanding of being and the historical. According to Heidegger, such a different understanding and way of being might emerge from thinking "the other beginning" in contrast to the "first beginning" that dominates Western metaphysics and history. Heidegger's text would thus be an attempt to recognize and respond to violence and its trauma rather than offering an apologetic valorization of violence that remains oblivious to its intrinsic pain.

Rereading Heidegger's Introduction to Metaphysics

Heidegger's *Introduction to Metaphysics* offers a different understanding of "agonistic thought" that contests its assimilation to a Social Darwinist model of the "struggle for existence," as Heidegger himself contends that competition and struggle are derivative modes of agon (the Greek word for "contest") rather than its defining feature.[11] Heidegger articulated in his interpretations of Heraclitus and Nietzsche how polemos is to be understood according to the question of being rather than in terms of human conflicts and attempts at domination. According to Hans Sluga, it is this difference between ontic and ontological violence that distinguishes Heidegger's reading of the polemos of Heraclitus from that of Carl Schmitt, where it is understood in the context of the difference between friend and enemy.[12] Whereas conflict is ontic conflict between competing humans for Schmitt, such that war and violence is always justified as the essence of the political, Heidegger interprets conflict ontic-ontologically as a question of being itself as well as of human existence. Heidegger could accordingly propose in his Nietzsche lectures of the late 1930s, challenging its fascist appropriation, that polemos does not justify but instead provides the basis for a critique of the self-assertion and struggle of egos and races. According to this reading, the agonistic or conflictual character of and the difference intrinsic to being can throw into question human conflicts and their motives. Human struggle, which always shows the dependency of those who struggle (*IM* 146), thus needs to be rethought in regard to the difference and agon of being itself. From this perspective, Nietzsche plays a crucial role for Heidegger as he also does for Schmitt. Nietzsche, whose philosophical works were distorted by National Socialism to support its biologism, racism, and will to power, also provides the basis for an alternative to and critique of the justification of the violence that humans do to one another. According to Heidegger, Nietzsche, as the victim of metaphysics, is also a witness to it (*IM* 28).

For Heidegger, Nietzsche is a witness precisely to its rupture and trauma. Recalling the critique of the priestly character in Nietzsche's *On the Genealogy of Morality* (I, 6),[13] it is the traumatic results of this violence that are concealed and deepened through a repetition that never heals the original wound. Not only is the pain not healed, but the treatment is worse than the disease (*GM*: I, 6). As opposed to the healing power of forgetting, Nietzsche argues that the pain is intensified by the way it is left unencountered and unquestioned, becoming ressentiment and hatred (*GM*: I, 10). This pain is only deepened in being reined in and tied to memory in order to shape a being that can and must remember for it to be calculable and controllable (*GM*: I, 1; II, 3). The National Socialist ideologue is, as a result, more akin to Nietzsche's sickly and life-denying priest, who in responding to pain does not respond at all and who poisons the wound in attempting to cure it, since he does not find meaning in himself but in the construction of and resentment against the enemy.

Heidegger's polemos should be interpreted as something that occurs to Dasein rather than as something that Dasein undertakes in competition with others to the extent that Heidegger's articulation of polemos suggests a "polemos without will" and without the "struggle for existence." Rather than constructing an agon between wills seeking to dominate and eliminate each other, Heidegger challenges human conflict and violence through the thought of a confrontation and polemos beyond violence and domination.[14] Heidegger described this conflict or strife beyond human will and action in the mid- and late 1930s as the strife of earth and world, an intercrossing strife that prevents the closure of being by a being called "man."[15] Polemos is then an originary strife, rather than a merely human agon or contest, that contests the reification of identity, whether of the self or community. It is precisely this reification that has been reinscribed in the interpretations of Nietzsche that Heidegger criticized during the late 1930s and early 1940s.

Heidegger argues that Nietzsche's thought is falsified if it is understood ethically as egoism or biologically as an assertion about race. His critique of Nietzsche's will to power a few years later, however, identifies the will with the assertion of the modern subject and its collapse in the enframing machination of technology (*Gestell*). For Heidegger, Nietzsche's rejection of system is not its overcoming but its completion in the total organization and—linking Nietzsche to Jünger—the mobilization of beings.[16] To this extent the various uses and misuses of Nietzsche's thought are not without their sources in Nietzsche's own works. The metaphysics of the will to power finds its historical culmination in the arbitrary subjectivism and totalitarian objectification of beings into objects of use; if not in Nietzsche himself then at least in the National Socialist employment of Nietzsche or—as Heidegger criticized it in 1941—"in the authentic Berlin interpretation of Nietzsche."[17] In contrast, Heidegger's "other thinking" strives for an "other attitude" in which man "does not calculate under the compulsion of utility and from the unrest of consumption."[18] Heidegger will accordingly attempt to think the event (*Ereignis*) of being beyond all agon and even polemos, as the primordial difference (*Austrag*) and confrontation (Auseinandersetzung) of the fourfold (*Geviert*) of gods and humans, world and earth.[19]

Heidegger's rethinking of agon offers points of departure for contesting the "contest of wills" model in which an entity called the will expresses itself, struggles for domination, and "triumphs" over others. Although Auseinandersetzung will only become more significant for Heidegger in the mid- and late 1930s,[20] it is already at work in his earliest project of a "hermeneutics of factical life."[21] John Caputo has argued that this project of a hermeneutics of facticity is tied to Heidegger's experiences of violence and conflict during the World War I.[22] It is also significant that Heidegger's early thinking of violence and conflict is bound to the possibility of a thinking of the agon and the polemos, which will unfold throughout his works as Streit, *Widerstreit*, and Auseinandersetzung. As we have begun to note, these terms do not only have the negative role of negation or setting limits in Heidegger's thought; they indicate something significant about factical life and its interpretive understanding in the early 1920s, and later the event, history, and thinking of being. These words indicate the originariness of violence and trauma in Heidegger's thought, that is, a thinking that provides a point of departure for encountering and questioning the constitutive role of violence and trauma in human existence without descending to a glorification of brutality and violence or absorption into a biologistic "struggle for existence." On the contrary, the interdependent difference and responsive conflict "for"—*für*, not *um*—the other articulated by Heidegger presents us with the means of destructuring the agon of wills, self-interest, and self-assertion that Levinas problematically finds in Heidegger.[23] Heidegger himself argues in the *Beiträge* that the thought of preservation of self and species belongs to the darkening of the world and destruction of the earth.[24]

The "setting-apart" of Auseinandersetzung does not end at the self—as if the self were a substance, subject, or essence—but the "self" is itself intrinsically uncanny or, more literally, "not at home" (*unheimlich*). According to Heidegger (and note that this translation employs the more literal "unhomely" rather than "uncanny"): "We understand the uncanny as that which throws one out of the 'canny,' that is, the homely, the accustomed, the usual, the unendangered. The unhomely does not allow us to be at home. Therein lies the overwhelming" (*IM* 115). Dasein is never purely "at home" with itself, it finds itself a stranger to itself. Self-appropriation cannot eliminate the finitude and uncanniness that constitutes Dasein's existence. The search for origins does not arrive at a pure homecoming that could eliminate the uncanniness of those origins. Dasein is overwhelmed by the violence of being and it responds with violence—not only against "being" but against other human beings and its own self, as violence always has an element of self-violence. History can be seen, according to Heidegger, as the unfolding of this violence by humans against being and against itself. Hence the human being is *to deinotaton* in the double sense of "the uncanniest of the uncanny" and "the most violent in the midst of violence" (*IM* 114–16). This violence remains hidden in the concern about human conflicts. Humans are the "violence doers" yet do not recognize the sources of their violence in their reactions to the overwhelming and violent character of being. The human being's violent response to being unfolds itself, in a passage that evokes Nietzsche's differentiation of will to power as self-creation and as domination, historically as either creativity (founding as opening) or domination (closing) through "machination" (*IM* 121–22).

The crucial question is not whether the violence of being and the counterviolence of humans occur but how they occur and how the human being, that is, that being that is a question for itself, responds. The issue is not merely one of "responsiveness," as there are diverse ways of responding, but of how to respond to violence and the trauma that it invokes. According to Heidegger, in a later essay, pain intrinsically joins as well as disjoins, gathers as well as disperses, it is difference itself interpreted as rift and separation.[25] Trauma is the lingering of pain that cannot be overcome, because it opens up a new world and comportment. Insofar as Heidegger articulates a constitutive rather than accidental pain, he is articulating the trauma that Elaine Scarry describes as "making and unmaking the world" in *The Body in Pain.*[26] Trauma does not only happen to a self that is one and the same before and after the traumatic event, as—akin to David Hume—there is no underlying unchanging self for Heidegger. Trauma opens up another world in which "everything has changed." In this sense, and despite her subjectivist language, Scarry's portrayal of pain cannot be set in diametrical opposition to Heidegger's as John Caputo and Andrew Mitchell have argued.[27] Because origins involve creative and/or destructive violence, they contain a trauma without recuperation. One can only begin to see "another beginning" precisely by confronting the "first beginning" in its upsurge, violence, and trauma.

Nietzsche depicted in *On the Genealogy of Morality* how the ascetic priest and his modern heirs deal with violence and trauma through reacting without adequately responding. The reified self does in fact begin as a response to trauma. Yet it is a response that repeats, reinscribes, and intensifies trauma insofar as it denies this world for an imaginary beyond (whether religious or political) devoid of conflict and suffering. Consequently, the cure is worse than the disease (*GM*: I, 6) and "poisons the wound" (*GM*: III, 15), as this pain is cultivated into revenge and resentment (*GM*: I, 7; III, 15). This intrigue and complicity of trauma and violence, of love and revenge, is nowhere more present for Nietzsche than the "gruesome paradox of a 'god on the cross,' that mystery of an inconceivable, final, extreme cruelty and self-crucifixion" (*GM*: I, 8). Nietzsche's *Genealogy* can thus be read as a genealogy of traumatic origins tracing the transformations of trauma and pain at the heart of present human practices and institutions.

Heidegger pursued an analogous genealogical strategy in his *Introduction to Metaphysics*, in which destructuring transformative repetition confronts habitual compulsive repetition. Instead of responding to violence and its trauma, the trauma is sublimated in Western metaphysics such that it is repeated and heightened. This unacknowledged "repression," besides bringing Heidegger into proximity with Freud,[28] informs the subsequent history of the West. Western onto-theology, including its culmination and fulfillment in technological modernity, is the repetition and intensification of an unacknowledged and poisoned traumatic wound. The counterviolence of humans against the violence of being—in creating artwork, political institutions, and other forms of thought and action—haunts those very beings as self-violence. In reacting violently to the violence of being, "Dasein commits the ultimate act of violence against itself" and Dasein "must indeed shatter against being in every act of violence" (*IM* 135).

Contrary to the assertion that Dasein can with heroic virility "master" its death, and consequently the interruptive trauma of existing, Heidegger already argued in *Being and Time* that Dasein cannot possess itself in its looming death but is, on the contrary, "shattered."[29] Heidegger repeats and transforms this claim in his *Introduction to Metaphysics*, where he describes how the human being, who responds to the violence of being through violence against beings, shatters on death (*IM* 121/168). Indeed, the violence-doing of Dasein does not master being but "*must* shatter against the excessive violence of being" (*IM* 124/173).[30] Dasein fails to overcome the trauma of its being done violence or being violated by what is its own violence. History accordingly shows for Heidegger both how humans exist out of violence and how violence is constitutive of that history. The question is then one of how to respond to this violence and its trauma. As I will argue in the conclusion, this response for Heidegger requires a transformation of our capacity to hear and see. It involves a revolution from self-assertion and the struggle for existence to the responsiveness of *Gelassenheit* as a letting that releases things, others, and being into their own, and therefore itself.

Heidegger depicts humans as the uncanniest and most violent of beings in his interpretation of Sophocles in the *Introduction to Metaphysics*.[31] This has been read by Caputo and others as a justification of violence. Nevertheless, it calls us to respond to violence and repeat it in such a way as to interrupt the compulsive identity of its repetition. Heidegger, at least by 1935, recognized the questionability of this violence—the violence of human conflicts that is rooted in a particular stance toward being. Heidegger thus claims that there are no origins without difference and violence. However, the recognition of this claim of constitutive or ontological violence can intimate another way of responding rather than being a celebration and endorsement of ontic or human violence. Indeed, Heidegger's discussions of the domination of power and the power of nonpower indicate that he is confronting the struggle for existence with a thought of being that undermines the endorsement of struggle and violence through the assertion of difference. This responsiveness is powerless rather than being another counterpower, and yet it is only powerlessness that can begin to undermine the dominance of power. This powerlessness is hence not passive recognition but responsiveness to the conflict of earth and world, which precludes the closure and systematic totality of metaphysics and modern politics. This "non"-power calls for a responsiveness that is precisely the encounter with the historicity of decision. Decision then, for Heidegger, needs to be understood in the context of the human response to being. Decision is not only a human occurrence, nor does it occur without responding to being, but it is the crossing of humans and being in which "history as such begins" (*IM* 84). Decision is not an empty activism nor is letting a mere passivism, as both the language of decision and of letting need to be considered from the perspective of the ethos (which for Heidegger means dwelling) of a fundamental responsivity. This responsiveness, of letting the thing show itself from out of itself, is not just to entities or things but also to their nontotalizable context and horizon, to the event of being itself. A responsiveness that is in one sense godless, and inhuman is ironically the condition of being responsive to gods and mortals, sky and earth in the openness, crossing, and between of the fourfold.

Conclusions

Various discourses justify, ignore, and excuse violence and its traumatic consequences in the name of various ideals, norms, and values that preclude any genuine confrontation, encounter, or response. Often, if we follow Nietzsche and Heidegger's reasoning, suffering is sublimated in an incessant repetition that only deepens the wound. This reifying pattern characterizes the perpetuation of "priestly power" in Nietzsche's *Genealogy*, as the violence of the present depends on the cruelty of the past and is codified as "good" in common life through tradition, habit, and custom. Could, however, individuation occur precisely in response to the uncanniness that cannot be mastered or overcome? Could being wounded call forth a reply that recognizes its wounded character, letting the wound appear as wound even in the case of the potentially irrevocable traumatic wound? Given the interpretive strategy of this chapter, and the importance of these questions, conflict (Widerstreit) and differentiating encounter (Auseinandersetzung) might well be central to the issue of trauma rather than accidental to it. Heidegger's thinking responds to the issue of trauma insofar as it articulates the possibility of a releasing responsiveness in the context of being shattered and broken-down as events that are not "secondary" to human existence but throw open its very significance.

Heidegger's approach, although it is not directly concerned with ethics or social theory, does not preclude the ethical and social dimensions of human existence. In *Being and Time*, the identity and difference of Dasein (being-there) is constituted in everyday being-with-one-another (*Miteinandersein*). This being-with (*Mitsein*) can be articulated through the concept of an agonistic responsivity or what he called an "interpretive setting-apart-from-each-other" (*verstehende Auseinandersetzung*).[32] This poses the question of whether responsiveness can be thought, as Levinas and Gadamer suggest, without the violence of difference and separating encounter, or whether it cannot be.[33] If the latter, then the limits and questionability of responsiveness itself can begin to be seen in the abyssal event and occurrence of interdependent differentiating conflict (Widerstreit). Such Streit is productively constitutive of logos and language through which humans attend or do not attend to things and each other. It indicates an agon or polemos without will or the self-assertion of the will.[34]

In contrast to domination, listening confrontation occurs out of the "between" and the abyssal divide, as the question and answer concerning the violence and uncanniness of the human. Heidegger explored in his *Introduction to Metaphysics* (in light of strategies comparable to Nietzsche's genealogy) this relational context or nexus of address and conflict, of logos and polemos. In *Wege zur Aussprache* (1937), Heidegger points out the prospect and risk of interpretive confrontation (verstehende Auseinandersetzung). He already used this expression in the early 1920s, and in both contexts it is concerned with a recognition of the other that does not forget the differentiation of the "between" or "with" linking and distancing self and other. This difference is understood as a conflict (Streit), not for the sake of strife but for understanding the other.[35] This is because difference (*Unterschied*), a difference that will no longer be spoken of in the language of violence in his postwar texts, is announced in hearing.[36] For Heidegger, "we can truly hear only when we are hearkening" (*IM* 99). We can only hearken when we are responsive to what is said. Responding, however, is something barely heard in the word "correspondence" (*IM* 95). This correspondence needs to be rethought from out of the context of being claimed and answering that claim. Through such answerability, Heidegger conveys another kind of hearing, even if we are inexperienced in such hearing and our ears overcome by the violence that thwarts responsive hearing and interpretive confrontation (*IM* 112). Although we can articulate this prospective hearing through Heidegger, it is also significant that the dangers of not hearing and not responding to suffering can be seen in Heidegger's own moral and political failures during the 1930s—the painful facticity of which evades being sublimated or overcome.[37]

Notes

I would like to thank Bettina Bergo, Robert Bernasconi, Kristen Brown Golden, and Jeanne Marie Kusina for their comments and criticisms of earlier presentations or drafts of this chapter.

1. See, for example, Jean-Luc Nancy, "The Being-with of the Being-There," in *Rethinking Facticity*, ed. François Raffoul and E. S. Nelson (Albany: State University of New York Press, 2008), chap. 3.

2. *GA* 27, *Einleitung in die Philosophie* (Frankfurt: Klostermann, 1996), 220.

3. A number of the chapters in Richard Polt and Gregory Fried, eds., *A Companion to Heidegger's Introduction to Metaphysics* (New Haven, CT: Yale University Press, 2001) take up these issues in a more serious way than previous polemical and apologetic works. Hereafter cited in the text as *Companion.*

4. References are to Martin Heidegger, *Introduction to Metaphysics*, trans. G. Fried and R. Polt (New Haven, CT: Yale University Press, 2000). Hereafter cited in the text as *IM.*

5. Heidegger articulated in *GA* 90, *Zu Ernst Jünger*, (Frankfurt: Klostermann, 2004) a deep critique of Jünger's totally mobilized subject. He interprets it as heightening rather than overcoming the modern subject. The fully self-consuming and mobilized subject remains a subject (*subjectum* as *Gestellte*) and is accordingly another affirmative expression of, rather than alternative to, modernity.

6. Compare the contributrions by Theodore Kisiel ("Heidegger's Philosophical Geopolitics in the Third Reich"), Frank Schalow ("At the Crossroads of Freedom: Ethics without Values"), and Hans Sluga ("'Conflict Is the Father of All Things': Heidegger's Polemical Conception of Politics") in *Companion*, concerning the political context of Heidegger's *Introduction to Metaphysics.*

7. Levinas criticized Heidegger's ostensive self-interested heroic decisionism and passive submission to an anonymous indifferent "Being" that reflects a totalitarian attitude as Tina Chanter notes in *Time, Death, and the Feminine* (Stanford, CA: Stanford University Press, 2001), 28. Habermas proposed a developmental schema in which Heidegger engaged in a heroic activist decisionism from *Being and Time* to his National Socialist engagement and, due to the conflicts and failures that this involved, thereafter turned to a passive resignation reflected in his language of *Gelassenheit* in "Work and Weltanschauung: The Heidegger-Controversy from a German Perspective," trans. John McCumber, *Critical Inquiry* 15 (Winter 1989): 431–56. Criticizing Heidegger's thought for its "heroic virility" is ironic given that National Socialist ideologues claimed that Heidegger fundamentally lacked it, since a philosophy concerned with angst, care, and death could only be for the decadent and weak. Compare Heidegger's response to their claims in 1941 (*GA* 49: 31–33).

8. Robert Bernasconi, "Levinas and the Struggle for Existence," in Addressing Levinas, ed. E. S. Nelson, A. Kapust, and K. Still, (Evanston, IL: Northwestern University Press, 2005), 176 and 171–73.

9. Levinas's most classic statement of being as war and ontology as violence can be found in the preface to *Totality and Infinity*, trans. Alphonso Lingis (Pittsburgh, PA: Duquesne University Press, 1969), 21–30.

10. See John Caputo, "Heidegger's Revolution: An Introduction to *An Introduction to Metaphysics*," in *Heidegger toward the Turn*, ed. J. Risser

(Albany: State University of New York Press, 1999); and John Caputo, "Thinking, Poetry and Pain," *Southern Journal of Philosophy* 28 (1989): 155–81.

11. See *IM* 74 and Martin Heidegger, *Parmenides* (Frankfurt: Klostermann, 1982), *GA* 54: 26–27 / *Parmenides*, trans. A. Schuwer and R. Rojcewicz (Bloomington: Indiana University Press, 1992), 18.

12. H. Sluga, "'Conflict Is the Father of All Things': Heidegger's Polemical Conception of Politics," in *Companion*, 214–16.

13. Friedrich Nietzsche, *On the Genealogy of Morality*, trans. M. Clarke and A. Swensen (Indianapolis, IN: Hackett, 1998), hereafter cited in the text as *GM*. The first number refers to the treatise, the second to the section.

14. Note both *IM* 47 and *GA* 69, *Die Geschichte des Seyns* (Frankfurt: Klostermann, 1989), 8.

15. See, for example, *GA* 69: 19.

16. *GA* 67, *Metaphysik und Nihilismus* (Frankfurt: Klostermann, 1999), 159.

17. *GA* 49, *Die Metaphysik des deutschen Idealismus (Schelling)* (Frankfurt: Klostermann, 1991), 122. Also compare Heidegger's comments on a statement by Adolf Hitler in which he argued that justification is reduced to usefulness for the collective in *GA* 66, *Besinnung* (Frankfurt: Klostermann, 1997), 122–23.

18. *GA* 51, *Grundbegriffe*, 2nd ed. (Frankfurt: Klostermann, 1991), 4–5 / *Basic Concepts*, trans. Gary Aylesworth (Bloomington: Indiana University Press), 4.

19. *GA* 67: 77.

20. Compare Charles Scott's remarks about Auseinandersetzung and hearing: "This *Auseinandersetzung* in its leonine aspect is also conditioned by an effort on [Heidegger's] part to hear the way in which philosophy presents things, to hear philosophical thinking in a way that is appropriate to it, and to give voice to his differences from it as he encounters it. This hearing aspect that conditions *Auseinandersetzung* constitutes a *Zwiesprache*" in "*Zuspiel* and *Entscheidung*: A Reading of Sections 81-82 in *Die Beiträge zur Philosophie*," *Philosophy Today* 41 (1997): 162–63.

21. Especially *GA* 61, *Phänomenologische Interpretationen zu Aristoteles*, 2nd ed. (Frankfurt: Klostermann, 1994), 2. See my account of Auseinandersetzung and its ethical import in "Ansprechen und Auseinandersetzung," *Existentia* 10 (2000): 113–22.

22. John Caputo, "Sorge and Kardia," in *Reading Heidegger from the Start*, ed. John van Buren and Theodore Kisiel (Albany: State University of New York Press 1994), 327–43.

23. See *GA* 27, 22–23, and 327 as well as Bernasconi's discussion of this question in "Levinas and the Struggle for Existence," 170–80.

24. *GA* 65, *Beiträge zur Philosophie: (Vom Ereignis)* (Frankfurt: Klostermann, 1989), 277 / *Contributions to Philosophy: From Enowning*, trans. P. Emad and K. Maly (Bloomington: Indiana University Press, 1999), 195.

25. Martin Heidegger, *Poetry, Language, Thought*, trans. A. Hofstadter (New York: Harper and Row, 1971), 204.

26. Elaine Scarry, *The Body in Pain* (Oxford: Oxford University Press, 1985). Andrew Mitchell argues, however, for the inadequacy of Scarry's account in relation to Heidegger in "Entering the World of Pain: Heidegger's Reading of Trakl," in *Telos* 150 (Spring 2010).

27. Mitchell, "Entering the World of Pain," and Caputo, "Thinking, Poetry and Pain," 180, fn.2, make this argument for opposite reasons; the former finds Scarry's subjectivizing and internalizing language problematic, and the latter contends that it is Heidegger who misses the reality of pain.

28. Uncanniness and repression are developed in a very different way in the writings of Sigmund Freud. For Freud, the uncanny (*Unheimlichkeit*) arises in seemingly accidental slips and dreams, showing that the ego's layered dynamism is not as settled and structured as the ego believes but that it is also uncanny, ungrounded, and homeless.

29. Martin Heidegger, *Sein und Zeit*, 16th ed. (Tübingen: Niemeyer Verlag, 1985), 385.

30. Heidegger confirms this point further in his *Contributions to Philosophy*, in which he discusses how the mastery and understanding of the subject is shattered in the anticipation of its own death in *Being and Time*: "But understanding of being is throughout just the opposite, nay even essentially other than making this understanding dependent upon human intention. How is being still to be made subjective at that place when what counts is the shattering of the subject?" (*GA* 65: 455-56/321).

31. Claire P. Geiman provides an insightful account of the issues involved in Heidegger's discussion of Sophocles in this work in "Heidegger's Antigones" in *Companion*, 182. On Heidegger's repeated encounters with Sophocles, also see Véronique Fóti, *Heidegger and the Poets* (Atlantic Highlands, NJ: Humanities Press, 1992).

32. Compare the early use of this expression in *GA* 61: 2 and the later use of it in *Wege zur Aussprache* in *GA* 13, *Aus der Erfahrung des Denkens* (Frankfurt: Klostermann, 1983), 15–21.

33. Gadamer suggested, for instance, that violence is not a consequence of Heidegger's account of understanding but rather is due to Heidegger's practice of "productive misuse" and his "lack of hermeneutical consciousness" in *Truth and Method* (New York: Continuum, 1989), 501.

34. For Levinas, an interest in being essentially reflects the "survival instinct," self-interest and being are at the root of violence in, for example, *Entre Nous: Thinking of the Other*, trans. M. B. Smith and B. Harshav (New

York: Columbia University Press, 1998), xii. Bernasconi has argued that Levinas's critique of Heidegger rests on a critique of the self-assertion of the will and ego in the "struggle for existence" (*Kampf ums Dasein*) in "Levinas and the Struggle for Existence," 171–73. Levinas's criticisms consequently presuppose an egoistic and biologistic interpretation of Heidegger that this chapter places into question. This issue is more complicated even in the case of Nietzsche, who saw himself as a critic of Darwin in interpreting life through the "will to power" rather than "survival of the fittest." Nietzsche rejected the Darwinist notion of a "struggle for life" in favor of the self-assertion of and "struggle for power" in *Twilight of the Idols*, trans. Richard Polt (Indianapolis, IN: Hackett, 1997), 59.

35. *GA* 13: 15–21.

36. Note the following reflections on hearing in Heidegger: *GA* 55 *Heraklit*, 3rd ed. (Frankfurt: Klostermann, 1994), 238–60; David M. Levin, *The Listening Self* (London: Routledge, 1989), chaps. 2 and 6; Scott, "*Zuspiel* and *Entscheidung*," 162–63; and Peter Trawny, *Heideggers Phänomenologie der Welt* (Freiburg: Alber, 1997), 90–96.

37. On facticity, and its ethical implications, see my chapter "Heidegger and the Ethics of Facticity," in *Rethinking Facticity*, ed. François Raffoul and E. S. Nelson (Albany: State University of New York Press, 2008), chap. 6.

Martin Heidegger Bibliography

Einführung in die Metaphysik, 4th ed. Tübingen: Niemeyer, 1976. / *Introduction to Metaphysics*, trans. G. Fried and R. Polt. New Haven, CT: Yale University Press, 2000.

Poetry, Language, Thought, trans. A. Hofstadter. New York: Harper and Row, 1971.

Gesamtausgabe. Frankfurt: Klostermann, 1976–ongoing.

GA 13: *Aus der Erfahrung des Denkens* (1983).

GA 27: *Einleitung in die Philosophie* (1996).

GA 49: *Die Metaphysik des deutschen Idealismus (Schelling)* (1991).

GA 51: *Grundbegriffe*, 2nd ed. (1991) / *Basic Concepts*, trans. Gary Aylesworth. Bloomington: Indiana University Press, 1993.

GA 54: *Parmenides* (1982) / *Parmenides*, trans. A. Schuwer and R. Rojcewicz. Bloomington: Indiana University Press, 1992.

GA 90: *Zu Ernst Jünger* 2004).

11

Trauma and Hysteria

A Tale of Passions and Reversal

Bettina Bergo

Introductory Remarks

This chapter studies the unfolding of Freud's concept of hysteria from its beginnings as heredity and as the effect of traumatic incidents,[1] through its displacements toward repressed ideas or fantasies, and identificatory conflicts. To trace this evolution, I discuss the innovations proposed by Freud's French mentor, neurologist Jean-Martin Charcot, in the study of hysteria. Charcot's innovations included two significant observations: first, that mind and body were so fundamentally intertwined that a psychological event could produce extraordinary neurological symptoms like long-term retention of urine, spontaneous hemorrhaging, and localized paralysis; and second, and against the entrenched convictions of the time, hysteria was a disorder that affected women and men.

As I examine the unfolding of Freud's explanation of hysteria, which led to his largely abandoning further investigation of the condition after World War I, I also describe the cultural milieu in which Freud worked and the philosophical and medical assumptions prevalent in his time concerning women, Jews, and peoples under colonial rule. While Freud's great discovery of psychic repression—and with this, a complex unconscious—established psychoanalysis as a discipline and the "talking cure" as its primary therapeutic strategy (over a host of nineteenth-century therapies from magneto-therapy to hypnosis), the evolution of Freud's discovery diminished the importance of trauma as the main cause of hysteria. Theoretical gains were thus accompanied by a certain loss. I will show how and why this occurred while remaining attentive to the categories, organized according to binary oppositions, prevalent in European and American conceptions of masculinity, femininity, Jews, and other "Others."

French Neurology Becomes a Scandal in Vienna: The Male Hysteric

Freud's version of the early history of psychoanalysis has an epic quality that obscures debates and hesitations. This is true for the absorption of traumatism into the disorder tentatively called "hystero-epilepsy" from the mid-eighteenth century. Following the work of the psychiatric clinicians, C. Lasègue et P. Briquet, hysteria had become an object of medical interest in Paris by the time Charcot delivered his first lectures on the condition in June 1870. Together with his "right-hand man" Désiré Bourneville, Charcot argued that hysteria should be separated from epilepsy and pointed out a distinctly affective component in the hysterics' convulsions—*réminiscences.*[2] World renowned as a neurologist by the 1880s, Charcot hosted the young Freud at his Salpêtrière clinic in the spring of 1886. Upon his return in October 1886, the young *Dozent* in neuropathology presented his findings to colleagues at the Viennese Society of Physicians. While Charcot's findings were controversial mainly for their insistence that hysteria was a single, neurological disorder, the question of who was affected by it did not much distress the French medical establishment. Charcot had documented cases of hysteria in women and men—notably Jewish men fleeing the outbreak of pogroms in Russia[3]—and Freud now proceeded to present the variety of symptoms observed at the Salpêtrière. He had begun to describe the etiology in Charcot's language when he was interrupted by a scandalized audience. The idea of male hysteria "was incredible," they erupted. In Freud's own narrative, "One of them, an old surgeon, actually broke out with the exclamation: 'But, my dear sir, how can you talk such nonsense? *Husteron* [*sic*] means the uterus. So how can a man be hysterical?'" As Freud noted years later, his surgeon colleague—who went unnamed but may have been Freud's Viennese teacher, Theodor Meynert—not only perpetuated myths like that of toxic organic vapors or the centuries-old wandering womb, both still believed to lead to hysteria when the uterus exuded pollutants or got "lodged in the throat."[4] The unnamed surgeon did not even know that the Greek source was the feminine *he hustera*, not the neuter *tò husteron*. The hoary physician did not even "know his Greek" and it took an outsider to explain the misunderstood malady and unmask the lack of classical culture in his colleague.[5]

Sander Gilman echoes Freud's story in an essay on the cultural elaboration of "the Jewish Psyche." This is part of a larger work on the nineteenth-century European cultural imagination, and the ways in which this imagination classified of Jews and other Others, notably women. Now, ideas about emotions, passions, and psychiatric disorders, moving out of France toward Austria and Germany, parallel the rise of urban Jewish populations in France, but especially in Berlin and Vienna,[6] as well as the emergence of "scientific" anti-Semitism in German-speaking cultural institutions (*JB* 61). This is important because, in the fin de siècle bourgeois society, notions of cultural decadence—discussed for some fifty years by philosophers from Nietzsche to de Gobineau—flowed together with the theme of species or racial degeneration, derived from hygienicists and from the social Darwinism of Herbert Spencer and Ernst Haeckel (author of the "recapitulation theory," dear to scientific racists and to Freud himself). Though not quite a full-blown ideology, Spencer's cosmic evolutionism extended Darwin's ideas to cultural and ethnic groups, legitimating the notion of a hierarchy of collective unconsciouses, each one conditioned by their "fitness for survival."[7] Zeev Sternhell, the eloquent student of French fascism, argues that "applied to society, Darwin's hypotheses no longer constitute a scientific theory, but become a philosophy, virtually a religion. . . . These new theories completely reject the traditional, mechanistic conception of man, which argues that behavior is commanded by rational choice."[8] They were not the only theories to impugn rational choice. Responses to materialist rationalism punctuate the Modern period, and psychoanalysis will prove an occasional ally in this.[9] Be that as it may, in the toxic confluence of nineteenth-century "sciences," appraisals of character, body, and cranial traits, intelligence and dispositions that belied inherited physical degeneracy or an ineptitude for survival, were attributed to the indigent, to non-Europeans, and preeminently to the recently expanded Jewish populations of Paris, Berlin and Vienna. Gilman writes:

> Freud's understanding as against the understanding of his time was that hysteria did not manifest itself as a disease of the "womb," but of the imagination. This did not absolve the female from being the group most at risk, however; for the idea of a pathological human imagination structurally replaced the image of the floating womb as the central etiology of hysteria. What was removed from the category of hysteria as Freud brought it back to Vienna was its insistence on another group, the Jews, which replaced the woman as essentially at risk. (*JB* 61)[10]

Since the Franco-Prussian War, scientific racism had insisted that Jewish men evinced a large number of the symptoms that would be attached to hysteria, from claudication or limping, to histrionic affect, dissimulation, and the more traditionally anti-Semitic.

"Conceptual Contagion" and the Pathological Imagination—of European Science

The displacements Gilman notes here were not definitive. Whether transferred from women to Jews, or women to Africans, children to other Others, type-based displacements are never complete and enduring, because the dynamic quality of symbolic abjections—whether effected by psychiatry, medicine, or literature—relies on a binarist logic that deploys concepts as positive and negative, as active (or virile) and passive (or effeminate), (racially) healthy and degenerate, ontologically "whole" or partial, diminished. Jews were depicted as possessed of an idiosyncratic intelligence, yet they were neurologically degenerate. Senegalese men, on the other hand, were physiologically healthy, but intellectually infantile.[11] European women were not healthy the way European men might be, by virtue of physical differences. More important, while frequently infantile, women were nonetheless not *eo ipso* degenerate.

Sander Gilman and Daniel Boyarin have done extensive work on the conflation of Jewishness and femininity in fin de siècle Europe.[12] There are two reasons why their observations are apposite here. The first reason concerns social imaginations: there is a striking symbolic contagion to nineteenth-century ideas on "femininity," with their association of weak passions, anxiety, fragile bodies, sexual continence, or promiscuity (*JB* 38–59, 145).[13] In the literary and popular imaginations, this contagion spreads to other Others, and espouses the additional characteristics attributed to them. These others become anxious, duplicitous, and otherwise intelligent, that is, possessed of a different logic.[14] I am not convinced that this is invariably tied to the pathologization of bodies for the purpose of exercising "bio-power" over them, as the argument has run since Foucault.[15] Georges Canguilhem's analysis of the development of social norms for health—and a host of other features from height to blood sugar levels to intelligence—is equally revealing.[16] Thus, in the 1880s, the French historian Ernest Renan pronounced, "the term Jew . . . denotes a special way of thinking and feeling."[17] A related example is found, some twenty years later, in the work of the Jewish classicist, Otto Weininger, *Sex and Character: An Investigation into First Principles.* The book was published in 1903 and grew popular following his suicide in the fall of that year.[18] Its diagnosis was unsurprising: European culture was decaying; degeneracy was manifest in Europe's growing effeminacy and "Judaicization." But in Weininger's argument, the redemption of Western civilization would take place through the liberation of women from themselves; that is, through the sublimation of sexuality, which turned women into instruments of masculine desire. The literature on the nature of women, Jews, and "ethnic" others was widespread. Whatever their degree of elaboration, the ideas were rooted in the supposition of cultural and biological decay, and in some cases, of cultural crisis.[19] The most powerful themes presented a vision of a debilitating effemination, conceived of as spreading like a disease to the military, the aristocracy, even to the working classes.[20] Freud's announcement that he had observed male hysterics in Paris thus produced a double scandal: the possibility of a pathological masculine psyche, and the fact that an "other" was holding a mirror up to Viennese society.

To be a Jewish neurologist in fin de siècle Vienna elicited questions of legitimacy by virtue of the same symbolic contagion. Gilman presents arguments that held that "Jews did different science," even exercised a different gaze, in the eyes of the nineteenth-century university. In Freud's case this worked like a gauntlet to be seized. Following the October 15 debacle, he produced before his colleagues, on November 26, a case of a Viennese male hysteric, Herr August P. In relating the etiology, we can see Freud explicitly psychologizing the condition of hysteria, moving it from the body—but also the imagination—to Bourneville and Charcot's trauma-based causality that conceived the mind-body relationship in a new way.

In a different time and place, the presentation of a male hysteric would have elicited little more than curiosity. By doing so in Vienna, Freud held up a mirror that reflected his colleagues' own anxieties about identity, cultural and economic change, as well as the growing instabilities to which syndicalism, women's and homosexuals' movements contributed. More important, Freud hobbled an ideological mainstay he scarcely imagined he was touching: the essentialist conception of masculine versus feminine characters and disorders. I will return to that point later. For now, I am concerned with the evolution of hysteria and trauma. When Freud returned to the University of Vienna with a male hysteric recently come under his care,[21] he was insisting on a battle that he would ultimately win: the psychologization of what was conceived of either as a physiological, a sort of "autoimmune" disorder (the noxious humors model for female bodies) or a religious aberrancy (akin to possession). Thus he was modifying a category in semantic flux under the pressure of scientific and political debates. That a neurological disorder proper to women by virtue of their fragile physiology could be detected in men demanded the rethinking of the sense of hysteria and the relationship between bodies, psyches, and time. As Gilman argues, hysteria went from being a disease of the body to being a disease of the imagination. This is true of course for the European imagination. But we know since the publication of the unexpurgated correspondence of Freud and Fliess that trauma—in the form of child abuse—persisted as a question for Freud in the etiology of hysteria, as also in obsessive compulsion. Daniel Boyarin reminds us that, even after he had announced he was abandoning his "neurotica," Freud continued to deliberate about the role of "seduction" in neurosis (*UC* 191 n.7; 198 n.33). In a word, trauma and memory remained together at the root of hysteria's complex etiology.

The Strange Career of Hysteria Etiology

While the literature on hysteria is vast and multiperspectival, Dianne Sadoff has scrupulously traced the evolution of hysteria as a disorder.[22] On her account, hysteria passes through three overlapping stages before Freud encounters it at the Salpêtrière clinic. The beginnings of her account are close to Canguilhem's genealogy of the science of pathology. In essence, as long as medicine conceived the body within an anatomical paradigm, hysteria was a disease of the womb in a body whose organs were the loci of scientific attention, compared to vessels apt to move north and south, organs influenced by each other through proximity (*SF* 71). This was particularly true of the womb. The anatomical theory was present through the Enlightenment, along with enduring suspicions about the demonic possession of certain women. As concern with disease focused more on the dynamic physiology of the tissues, hysteria changed into a disease of imbalanced animal spirits in a body-container, where the locus of pathology was tissular (*SF* 59). As neurology and reflex theory gained in prestige, hysteria had its third avatar and became a disease of the nervous system and neural irritations. The womb was said to be connected by nerves and veins to every other bodily organ; it influenced in a polymorphic fashion other organs therein.

In the mid-nineteenth century, the English neurologist and near contemporary of Charcot, Thomas Laycock, insisted that this curious female "wiring" accounted for women's "feminine virtues" and passions. Anticipating the social imaginary of the fin de siècle, Laycock set up a Manichean inventory, including: "compassion, kindness, piety, honesty, sincerity, constancy" (*SF* 68) and, on the other hand, "religious enthusiasm, erotomania, nymphomania, monomania, rage, jealousy, craftiness, and cunning" (*SF* 68). The uterus had become, under the neuropathological reductions, the seat of reflex action in women. Sadoff concludes, "When irritated . . . uterus or ovaries relayed reflex irritation to other parts of the body,[23] causing attacks, convulsions, insensibilities, and paralyses" (*SF* 66). At each point in this evolution—which follows no simple linear succession—female reproductive organs were the primary entity and force in hysterical etiology.

It should come as no surprise, then, that Freud's Viennese professors would identify female virtue, passions, and vices with the uterus—whether it floated, released humors, or radiated neural irritation. Only when a significant part of the pathological field could be shifted from physiology to psychology could there be a properly masculine hysteria. That shift began in the psychiatry of revolutionary France with Philippe Pinel and his innovative typology of "manias without delirium." In France, it proceeded through Esquirol, who pursued the relationships between reason and mania in institutionalized subjects and as a hybrid of psychological and "moral" elements in which the burden of moral responsibility rested on the internee. When the pathological field reached Charcot and Bourneville, they moved it further toward the psychological.[24] In this complex displacement, which involved changes in the symbolic meaning of the male and female bodies, the moment that sexuality was implicated, sexual "seduction" was as well.[25] All of this played itself out against the background of "genetic inheritance," which held certain populations—Jews notably—as more liable to hysteria than others. Within France and Germany in the 1880s (*UC* 209), heredity theory amounted to a multilayered racism that targeted those "present" (Jews) and those "absent" (Africans, Asians).

Concerned to distinguish epilepsy from what was called "hysteroepilepsy," and to determine the relationship between localized losses of feeling (notably around the ovaries) and inexplicable paralyses, Charcot's investigations into the perplexities of hysteria spanned some twenty-two years (1872–94).[26] While his diagnostics moved from the neurophysiology of "hysterical ischiuria" (retention of urine), seizures, and "ovaria" (traced to neurocerebral causes), to traumatism (affecting the illocalizable "psychic apparatus"),[27] Charcot ultimately insisted on three things in the matter of hysteria. First, the disease was a single physiological reality, though he was given to referring to its most visible symptoms in a rather Catholic vernacular as "stigmata,"[28] thereby mythologizing them. Charcot developed a precise taxonomy of four phases, as well as fourteen ancillary types of hysteria (*VC* 177 n.4). Second, Charcot argued for a certain duplicity—a histrionicism—on the part of the hysteric in the mise en scène of her or his hallucinations (in the third and most unique phase of hystero-epileptic attacks). Finally, Charcot maintained that the predisposing factor in hysteria was the neurological degeneration that ran in "psychopathic families." The importance of such a predisposition became manifest in the disease's precipitating causes: physical traumata, including everything from an acute fright to railroad accidents, falls or war injuries (*TSP* 19–20), which produced paralyses and contractures, and psychological traumata, consisting of a different set of symptoms (ovaraesthesis, retention of urine, vomiting). Surreptitiously, a dualistic perspective insinuated itself into Charcot's causal explanation. Though he observed in the hysteric a repeating dissociation of consciousness into pathological and "normal" states (*VC* 176)[29]—perhaps anticipating Freud's speculation on the dissociation of life and death drives subsequent to traumatizing events—Charcot steadfastly maintained that *la chose génitale* lay at the root of the disease, in women and presumably in all hysterics. In 1886, already a decade after his separation from Bourneville and his theory of *reviviscence*, Freud reports that Charcot whispered to a gynecologist colleague at a dinner party, "but in these cases [of hysteria] it is always a genital thing, always!"—"toujours la chose génitale, toujours" (*SF* 58). Yet the "genital thing" remains sufficiently ambiguous to connote sexual fixation, dysfunction, or again rape.

Psychologizing Hysteria and Sexuality

It may be that for Charcot la chose génitale was an ironic (or prudish) way of bundling sexual "abnormalities" together with women and other ambiguously gendered groups. The condescending motif of sexual frustration lacked etiological significance however, if one failed to acknowledge a neurological predisposition and a traumatic trigger of some kind. More interesting is that a part of the etiological difficulties hysteria posed lay in the tension between the posited unitary system of hysteria and the fragmentation of consciousness it provoked (*VC* 174–75). Expanding an insight of his predecessor Pinel, to the effect that consciousness had a capacity for multiple states or divisions over time, Charcot observed that mental illness had a periodicity, and the mentally ill—here, the hysteric—was in no way consistently symptomatic. To demonstrate this, Charcot (and later, the young Freud, collaborator of Breuer) provoked, in his *Policliniques*, symptoms of hemiplegia in patients under hypnosis. His aim was to show that if one could alter a patient's state of consciousness, then one might induce or remove hysterical symptoms by way of mild physical traumatism and suggestion. The ultimate object was to construct a bridge between the dissociative mental states of spontaneous hysterical attacks and those that could be precipitated in the clinic and then utilized as a therapy.

Thus, in a particularly dramatic session, published in 1888, translated and annotated by Freud upon Charcot's death in 1894, the neuropathologist produced a case of "psychic paralysis by autosuggestion":

> Mr. Charcot, to the subject: "Raise your arms in the air; place your hands on your head." The subject executes these various movements with ease, as she would also do in a waking state. At that moment, Mr. Charcot delivers, using his closed fist, a blow of very moderate intensity to her left shoulder, whereupon the upper left member becomes flaccid, dangling, and inert, absolutely paralyzed to voluntary movement. And, at the same time, in this member, which earlier possessed sensation, we note the absolute and total loss of cutaneous and deep sensitivity, the absolute loss of any notion of muscular orientation. (*TSP* 19)

Charcot continued his demonstration, extending the induced paralysis to the left leg and creating a transient hemiplegia. His explanation—as recorded by a young Freud so sufficiently astonished as to employ three times the adjective "absolute"—unfolded as follows:

> I will not, at present, insist upon the theory of the production of these paralyses, rather I refer you to the details I have given in this regard in my third volume. I would simply recall that the somnambulized subject [that is, under *la grande hypnose*] is in a special mental state, particularly favorable to suggestions. In sum, what has happened is this, to my mind: I struck the shoulder lightly. This light traumatism, this local shock, sufficed in a nervous subject *especially predisposed*, to produce throughout the full extent of the limb a feeling [*sentiment*] of numbness, heaviness, and the indication of paralysis; through the mechanism of autosuggestion, this rudimentary paralysis has rapidly become a real paralysis. It is at the seat of psychic operations, in the cerebral cortex in other words, that the phenomenon evidently takes place. The idea of movement is already the movement in the process of enactment; the idea of the absence of movement is already, if it is strong enough, the realized motor paralysis; all this is in perfect conformity to the data of the new psychology. (*TSP* 19–20; emphasis added)

Though Charcot's conception of sexuality was narrower than Freud's would be, and despite his observations about the histrionics and mimeticism of his hysterics, he had induced traumatism in all its phenomenality under a controlled alteration of consciousness and body. His observation that "the idea of movement is [neurologically and cerebrally] already the movement in the process of enactment" gave a Leibnizian twist to the Cartesian mind-body separation. It is as though he demonstrated the presence and efficacy of that sensuous gamut Leibniz had termed *les petites perceptions*, and evinced their sway in the absence of any conscious attention to them. If the idea of movement, conscious or semiconscious, is already the movement itself, then body and mind are not simply separate entities, they are interchangeable states. This was a contribution quite different from his impish observation about la chose génitale, and Freud extended it even as he abandoned Charcot's nineteenth-century supposition of inherited neurological predispositions to mental illness by virtue of (ethnic) degeneracy.[30]

Charcot thus paved the way for Freud's subsequent disavowal of rigid distinctions between "real" and "imaginary" anguish, because Charcot first psychologized traumatism by refusing to hold the physiological and the psychological, the real and the imaginary, separated. It was Freud's wager that if the patient himself or herself could discern the causative event, in a more conscious state of mind than that under hypnosis, the lessening of the traumatism might prove to be enduring. Initially, then, the reality of male and female hysterics was undeniable, even when their symptoms followed different patterns. The profundity of this mind-body imbrication opened new problems. Charcot's provocative insight—"the idea of movement is already the movement in the process of enactment"—led to two parallel etiologies for hysteria. On the one side, traumatic accidents such as the train wrecks that produced "railroad spine," or war injuries; on the other side, traumatic incidents such as a shock, sudden fright, or the actualization of a lost memory through association with an occurent mental image or representation.[31]

In pursuing Charcot's psychophysical imbrication, Freud loosened the mythico-physiological dimensions of hysteria until he had renounced even the former's techniques of *grande* and *petite hypnose*—not to mention his metallo- and magneto-therapies. Upon returning to Vienna, Freud argued that hysteria was more traumatism than disease (i.e., the psychological effect of a traumatic event). Thus, it was neither a disorder of wombs, of neurological reflex action, nor of degenerate races (*UC* 208, 209 n.75). Nevertheless, Freud long preserved Charcot's hysterogenic trigger points, the principal one of which was situated in the area of the ovaries in women—and in a similar place in men, despite their lack of these organs.[32] Indeed, as his 1886 narrative suggests, authentic traumatism required that violent or disruptive events be cumulative.[33] Later, following his work with Breuer, hysteria gravitated more toward Bourneville's characterization, as a disease of memories. In Freud, however, the reminiscence amounted to a resemblance between events whose incipience might not have been traumatic, and an ongoing event that revealed the sexual meaning of the earlier ones, endowing them suddenly with an affective force and ideational form that overwhelmed the psyche, producing traumatism retroactively.

The evolution of Freud's theory of trauma-induced hysteria, even as it remained faithful to Charcot's observations, effectively went beyond claims about la chose génitale or the histrionics of effeminate Others.[34] Herr August, Freud's male hysteric, suffered from the anaesthesias, paralyses, and dissociative states typical of his diagnosed condition. For Freud, these symptoms were precipitated by events in Herr August's life, which were progressively consolidated into traumatism. Repetition, microtrauma, and anxiety became the psychological core of hysteria's etiology. Some seven years before he set forth his psychoanalysis, and speaking still as a neurologist, Freud proclaimed:

> When, on October 15th, I had the honour of claiming your attention to a short report on Charcot's recent work in the field of male hysteria, I was challenged by my respected teacher, Hofrat Professor Meynert, to present before the society some cases in which the somatic indications of hysteria—the 'hysterical stigmata' by which

> Charcot characterizes this neurosis—could be observed. . . . I am meeting this challenge to-day . . . as far as the clinical material at my disposal permits. (*SE* I, 24–25)

Following his October presentation, the senior physicians of the General Hospital of Vienna had refused to allow Freud to use their material—so energetic was their resistance to, and in some cases their identification with, him.

Freud continued, "[A]s far as the clinical material . . . permits, by presenting before you an hysterical man, who exhibits the symptom of hemi-anæsthesia to . . . the highest degree" (*SE* I, 25).

What caused the hemi-anaesthesia and, as Freud notes afterward, Herr August's convulsions? Freud described his patient's impoverished family background, his parents, and his siblings. Then he noted the catalyzing traumatic incident. Three years earlier, "his brother threatened to stab him and ran at him with a knife" (*SE* I, 26). This incident was followed by a "fresh agitation": a woman accused Herr August of theft (*SE* I, 26). As a result of repeated distresses, "our patient exhibits, both spontaneously and on pressure, painful areas on what is otherwise the insensitive side of his body—what are known as 'hysterogenic zones.' . . . Thus . . ." and note the pronounced alteration in language, mobilizing the legitimacy of Charcotian neurology: "the trigeminal nerve, whose terminal branches . . . are sensitive to pressure, is the seat of a hysterogenic zone . . . also a narrow area in the left medial cervical fossa" . . . as well as "the left spermatic cord [is] very sensitive to pain . . . into the abdominal cavity to the area which in women is so often the site of 'ovaralgia'" (*SE* I, 30–31).

In the November talk, we hear clearly Charcot's imbricated registers of psychological and physiological symptomatology for hysteria. Though Freud had not yet explored what Bertha Pappenheim would call his "talking cure," his account combined an extended exploration of family background, coupled with anxiety-producing incidents. At this stage, the purely physiological explanation seems the more mythic of the components, as though he were following Charcot while neglecting the latter's "autosuggestion" hypothesis. Thus, Freud traced the hysterical stigmata on Herr August's body. He located the physiological seat in the nervous system and around the reproductive organs. He recalled Charcot's highly specific "hysterogenic zones," such that Herr August felt pain around the spermatic cord, which, in women, is the usual site for hysterical "ovaralgia."[35] Given his trauma-based explanation, and the Charcot-Freud endeavor to surpass simpler mind-body parallelisms, the hysterical male body proved analogous to the pathological female body. More than any neglect of uterine neurophysiology, it was this structural analogy that scandalized the Viennese physicians. Moreover, that it was presented to them by a Jew—a member of a group that the then popular "science" situated in the catch-22 of being alternately lustful, or effeminated vectors of syphilis through circumcision, or hysterical and asthenic (*FRG* 66–71; *JB* 53–56; *UC* 212, 230)—may have made Freud's observations seem more like censure than innovation.

Later, when through their collaborative practice Freud and Breuer inflected the psychological characterization of hysteria toward cumulative traumatism and analogical ties between events, saying, "the hysteric suffers from reminiscence," the separation between traumatic accident and traumatic incident seemed to deepen. We know now from his correspondence with Fliess that Freud encountered, between 1895 and 1898, events that he could not write up as cases. If his self-analysis of 1896–97 reveals the outlines of a theory of infantile sexuality and seduction (*TSP* 25), including his own (*UC* 200), the evidence of what he termed "precocious sexual traumatism" in 1896 was overwhelming. In a letter dated December 22, 1897, Freud reports one scene recalled by a patient from her third year of life. The child's mother is busy and she is listening.

> [The] father is one of those men who stab women, who require, erotically, bloody wounds. When she was two years old, he brutally deflowered her and transmitted to her a gonorrhea such that her life was in danger. . . . The mother, *now*, finds herself in the room and screams: "Bastard, criminal! What do you want of me? I won't do it! Who do you think you are with?" Then, with one hand she tears at her clothes and with the other grasps them close to her body, which produces a strange impression. Thereafter, disfigured by rage, she stares at a point in the room, covers her genitals with one hand and with the other hand thrusts something away from herself. Then she raises both hands, claws and bites in the air. While screaming and swearing, she leans backward, again covers her genitals with one hand, then falls forward, her head almost touching the ground, and finally falls backwards, gently, to the floor. . . . What most struck the child was the scene in which the mother is standing and leaning forward. She noticed that her toes were strongly turned inward. When the child was six or seven months old (!), the mother was in her bed, almost out of blood following an injury inflicted by the father. At age sixteen, she again sees her mother bleeding from the uterus (carcinoma), and this marks the beginning of her neurosis. (*TSP* 27–28)[36]

We see here how hysteria arises as "the *precipitate* of a reminiscence," even if Freud used this neutralized, chemical, literary metaphor as late as 1910.[37] The persistence of Charcot's insight is significant: given the specificity of the circumstances noted, the truth of the child's narrative is at least partly demonstrable. More important, however, is that in such cumulative traumatism, it is otiose to try to set fantasy apart from reality. It is not that Freud wanted ultimately to cover up the depravity of Viennese or European bourgeois society in his day. He was aware of sexual violence; while in Paris, he had attended Paul Brouardel's public autopsies at the city morgue, and he kept in his personal library copies of then popular works on "*attentats sexuels*."[38] Rather, "reminiscence" means that events become traumata when, by dint of their connection with a new event that proves analogous or otherwise illustrative (standing in as a metaphor), the sense they acquire restructures the psychic aggregate of memories, whether these entail representations or a reimagining of earlier experiences. The verifiability of these experiences is less important than the power their subsequently revealed sense takes on.

I would thus argue that hysteria in the sufferer proves to be more than Gilman's ambiguous "imagination." Understood in light of reminiscence, and suggesting an unconscious censoring function in the process, Freud's conception of hysteria as traumatism pursued Charcot's mind-body imbrication to its psychological conclusion,[39] as two inseparable dimensions of a single entity. Whether a forgotten or a censored memory returns as a somatic symptom or as a psychic one, say as an idée fixe, the trauma will have been an incident of some sort. But that incident will carry the kind of power hitherto attributed only to major accidents. This means that catastrophic accidents such as railroad spine or war wounds lose some of their significance for hysteria in the wake of what psychoanalyst Didier Anzieu has called "precocious sexual traumatism." We see this clearly when Freud insisted that war wounds can mobilize sufficient narcissistic attention in the patient to avoid becoming hysterogenic.[40] In the process of exploring the differences between war wounds and war (and other) traumatism—while still acknowledging the psychic costs to the victim of the "interfantasmatic" dimension of a passive child confronting a seducing (perverse) adult (*TSP* 25)—Freud distanced himself from masculine hysteria even while recognizing that many of his obsessive compulsive patients (generally male) also reported precocious sexual traumatism (*TSP* 25). An extended analysis could show that the obsessive compulsive disorder shared many symptoms with hysteria, but it was never so closely aligned with femininity or *Entmannung*, effemination. Daniel Boyarin has done some of this work.

During his active correspondence with Fliess in the 1890s, Freud saw male hysterics as patients. In his 1886 "The Ætiology of Hysteria," six out of the eighteen cases studied are male (*UC* 194). He was quite fond of one, whose fantasies he found impinging on his own in the dream of "Irma's injection."[41] Juliet Mitchell and Boyarin are convinced that Freud dropped the male in male hysterics as he unfolded the oedipal schema of psychic development, ignoring both sibling conflicts (Mitchell) and those family dramas that "gendered" children, whatever their "sex" was (Boyarin).

What is important is that the unconscious, whose conceptual beginnings go back at least to Charcot's states of consciousness, takes its fully Freudian form thanks to the repression and return of memories liable to become traumatisms (bodily or psychic symptoms). With Freud's psychological unconscious—which remains a bodily one,[42] though that proves difficult to think without sliding back into earlier parallelist conceptions—comes the confirmation of a larger philosophical apprehension that humans are ultimately incapable of exerting sustained control over their passions.[43] Dynamically and genetically, both of Freud's topologies turn on consciousness experiencing itself, and recollecting itself, as passive in itself and emerging out of itself. It must experience itself as subjected to forces whose immanence is nevertheless so alien that internal-external, intrinsic-extrinsic distinctions lose much of their meaning. It may actively represent these, verbally or through images, but they may well prove independent of intentional consciousness and subvert it.

Thus the career of hysteria—considered essentially as traumatism, and later approximated to neuroses of narcissistic libido investments[44]—led unexpectedly to a philosophical insight as compelling as the philosopher Kierkegaard's paradoxical "concept of anxiety" (1844). For Kierkegaard, anxiety was a cause and an effect. It was traumatism and trauma; the sign and efficient cause of the divided self. It was also the emblem of the spirituality of European culture by contrast with the Greeks: anxiety, not rationality alone, was the modality through which one realized, better, felt that one was simultaneously able to act (i.e., free) and standing under an obligation or imperative, which one only affirmed by transgressing it. For Kierkegaard, this was the psychological circle that conditioned our sense of fallenness. By 1919, Freud too would conceive anxiety less as a symptom of hysteria or repression than as an implicit sense of conflict, we might say a fallenness, with a law (of the father) but without moral wrong.

Freud's "Copernican Revolution" and the Transformation of the Passions

The arguments for hysteria as the effect of a trauma, and for traumatism (as trauma's psychological effect) together exemplify our loss of control over our passions. This dealt a severe blow to the heritage of philosophical, and popular, autonomy. Freud was anything but alone in effecting this blow, which had also been delivered by Nietzsche, Weber, Wittgenstein, and others. These metaphoric coups reinforced, dialectically, pre-Enlightenment thought according to which the passions indicated a suffering that demonstrated the inevitable heteronomy of human reason. Foucault's *History of Sexuality* and psychiatrist Gladys Swain's *Dialogue avec l'insensé* remind us that the talking cure replaced and enlarged the religious confession.[45] It also edged into the space of more philosophical schemes for the cultivation of passions into virtues. From variations on the Stoic ideal of *ataraxia*, or implacable mind, to Kant's Categorical Imperative, the Enlightenment philosophical compass pointed toward rational autonomy and the cultivation of a good will. Rationality should guide the passions into peaceable coexistence and, with that, procure responsible freedom for men. If this could not be so, then the promise of the Enlightenment, at least as a beautiful human life, was undermined—and with it, important aspects of its social optimism. It is no accident that Freud would speak of his conception of the unconscious as bringing about a third "Copernican Revolution" (after Copernicus and Darwin).[46] This third revolution dislodged the vision of humans, autonomous in their power of reason to master, and guide, their passions toward virtue. Nevertheless, we should recognize the converse: psychoanalysis also proposed a new *therapeia* for the human condition that it brought to light. And certain of his writings show that Freud conceived the possibility of a "heroic," transformative development, possible for certain humans—illustrated by Leonardo and the Moses of Michelangelo.[47]

A Deathbed Confession: The Hysterical Mentor

The episode of Herr August, Freud's male hysteric, has a canonic ending that we know well: the burgeoning of psychoanalytic practice and the ramifying conceptual bases for reading the bodies and discourses that psychoanalysis gave us. But the episode has a more ironic ending as well. This other ending illustrates the conceptual contagion discussed earlier. Freud's teacher of neuropsychiatry, Theodor Meynert, whose work and personality was so significant to him that Freud would exclaim: "I had been struck [by Meynert's research] while I was still a student."[48] This same Meynert, who readily joined the embittered attack on Freud on October 15, 1886, made a deathbed confession to him years later: "You know," sighed Meynert, "I was always one of the clearest cases of male hysteria." Identification replaced hostility—although hostility had masked identification, complicated by Meynert's aversion to Jewish students and what they represented to him. Freud reported this in 1900, at the time of his *Interpretation of Dreams.*[49]

Hysteria's Decline (and Rebirth), and the Return of the Gendered Logic of the Passions

This story should be longer.[50] Juliet Mitchell points out that "the apparent disappearance of hysteria from Western society is bound up with the advent of the psychoanalysis which it inaugurated. The period of Freud's manifest [concern with] hysteria was over by around 1900. 'Thinking,' and speaking the very theory and practice of psychoanalysis, had replaced it."[51] This is not entirely true. A number of psychologists today, including Judith Herman, have questioned the actual disappearance of what is often still called "conversion hysteria," with its battery of somatic symptoms (anaesthesia, urine retention, trigger points on the body, contractions, clownism).[52] Hysteria lost interest because Charcot never found its presumptive neurocerebral foyer; in Freud's time it grew less and less clear whether the condition actually covered a single mental illness or several conditions. But traumatism—with its diversity of physical and psychological symptoms—has not disappeared; it attests to the kind of mind-body imbrication that alone could account for the multiplicity of sources and signs of trauma. Indeed it is the uniqueness by which individuals struggle to restructure their lives in the wake of trauma that poses contemporary psychology its quandaries about traumatism: efforts at classifying it as a pathology have produced a host of proliferating disorders—some treatable, others not so. We should keep in mind that Charcot had argued that hysteria was incurable, given the inheritance of predisposition to it. The most important of these is of course post-traumatic stress disorder (PTSD). On the other hand, the decline of certain symptoms—from mutism to religious visions to ovaralgia—once considered proper to hysteria, may be credited to changes in the cultural imagination concerning the psyches of women, persons gendered effeminate, and different "races."[53] Moreover, while sexual dysfunction and aggression remain sources of traumatism, the impact of wars and the resemblance between hysterical symptoms—"shell shock" and PTSD—argues clearly that Anzieu's sexual "incidents" cannot just replace "accidents" in the etiology of traumatism.[54] Freud came to a comparable conclusion in 1918. But it is imperative to rethink what kind of logic one engages when speaking, simply, of cause here, as traumatism is neither simply a physiological nor a psychological "disorder."

Now we find ourselves facing a double story. First, the philosophico-cultural one. It was the scandal of hysterical men—victims of trauma—that brought to light a cultural imagination for which reason, passions, and bodies were approached according to varying arrangements of those mythic binaries mentioned earlier, including the polarity of normal versus pathological, the active versus the passive, the racially healthy versus racial or familial degeneracy, the eros of entangled immanence versus that of a spiritualized transcendence. That such essences and their related predicates could be so variously paired together was thanks to the stabilizing, symbolic logic typical of nineteenth-century medicine and psychiatry: a norm was interpolated, like an unconscious *petitio principii*, on the basis of a selected, local average; an (aestheticized) ideal was posited as the *definiens* to which objects to be defined stood, distant or close by, but not in themselves. The curious thing is that fin de siècle intellectuals took these binaries and their intersections so for granted that they did not write extensively about them—with the exceptions of Weininger and others—until aspects of the scientific *engrenage* were threatened.[55]

The second story, the psychological one, is more familiar, and perhaps more complex. Freud's investigation of hysteria motivated him to search for clear causes and grounds—trauma, suffering; later repression in emotional conflict replaced the racist "family disposition." With this shift he elaboated an unconscious different both from its German depictions from Schopenhauer to Hartmann, and from Charcot's; for Freud's unconscious was a metaphoric "site" invested—at different "levels" and with different quotients of energy whose "translations" included ideas, names and signifiers, and affects. The curious notion of unconscious affects denoted modes of embodying conflict—in anxiety, rage, guilt. Unconscious passions required the complexification of psychic causality and economy. Although Freud, following Charcot, moved hysteria toward the psychological account that revised the sense of philosophy's mind-body separation,[56] though he challenged philosophical and physiological conceptions of passions as the destruction of autonomy, Freud also pulled hysteria toward both precocious erotic "incidents" and repressed fantasies. The displacement toward repressed fantasies accelerated with Freud's second theory, that of oedipal development (*MM* 75), and his findings among the *Kriegsneurotiker* and their affective conflicts of loyalty. Yet Freud's "Oedipus" also structured a certain type of fantasy.

As we know, around 1910 the theory of oedipal development opened a new topos of the unconscious, different from the primary process–secondary process model of the earlier work and more explicitly "recapitulative" (family development recapitulates sociogenesis, the infant recapitulates phylogenetic-species moments). With oedipal development, the traumatic incident etiology of hysteria seemed to him ultimately to hold more explanatory power for *Zwangsneurose*, like the obsessions of the "Rat Man."[57] Concomitantly, the question of hysteria lost interest for him, especially after he published his notes on the interrupted treatment of "Dora." J. M. Masson has ventured one explanation for this: Freud believed he could not publish the case studies he had shared with his friend, Wilhelm Fliess. However, while he speaks clearly of their unsettling sexual content, Masson neglects the ties between hysteria, race theories, and degeneration. In the Fliess correspondence Freud spoke at length of male hysteria—notably of one "E" with whom Freud found that he shared certain symptoms[58]—but the cases he wrote up after 1897 were of women. For reasons ranging from the dissolution of the Freud-Fliess intimacy to a desire to avoid the insidiousness of racialization, the condition of hysteria returned to its place of privilege as a feminine pathology.

After publishing the Dora case (whose initial title was to be "Dreams and Hysteria"), Freud abandoned the field to its feminine and infantile poles as massively disordered passions and energies, precipitated by an event that restructured the conflictual, erotic dimension of forgotten incidents. In the popular reception of his work, the hysteric again became a "sensitive" individual (feminine, effeminated, or otherwise deteriorated by inherited syphilis or another degenerative condition) (*FRG* 114ff; 174), overwhelmed by adult sexuality. Though I would not deny Freud's insight into conflicts immanent to the ego in the etiology of hysteria, I agree with Judith Herman that the meaning of hysteria as a disorder of trauma and abjection could only resurface after 1918, when "the reality of psychological trauma was forced upon public consciousness once again by the catastrophe of the first [*sic*] World War" (*TR* 20). Perversely, when it did resurface it did not present itself as hysteria. We might see this as a certain refinement, but there is a surprising, repressive aspect to the new hysteria. It largely became a physically caused shell shock—an explanation that Freud himself challenged for its weak explanatory power (some soldiers suffered no wounds yet developed hysterical symptoms). Despite the revival of physical trauma as the scientific etiology, popular treatment strategies took on a disconcerting appearance. They concentrated less on "talking cures" than on Philippe Pinel's punitive "moral" measures from the early nineteenth-century psychiatric wards: "shaming, threats and punishment" (*TR* 21).

Concluding Remarks

In this account, the adherence of traumatism to sexual incidents befalling women or children left its disorder, hysteria, as at least an effeminate (or *Untätigkeits-*, idleness) neurosis. The advantages accruing to Freud's incident (trauma) etiology included the impetus to expand his dynamic, economic, and topological unconscious. The difficulties included a certain decomplexification of gendering as a process, and the ultimate refeminization of hysteria. Yet trauma-based hysteria in men was the very problem Freud had first addressed with youthful temerity. Additionally, while in 1886 repeated traumata had produced Herr August's ultimate breakdown—whose principal symptoms included intense anxiety, disorientation, and anaesthesia—sexual incidents or abuse were not among his recollections, any more than they seem to be part of "E"'s hysterical etiology. And from his self-analysis we know that male hysteria had a different relationship to the precocious sexual incident etiology Freud noted in women patients.

I argued earlier that Freud's initial insistence that hysteria was found in men and women, and that it was the effect of trauma, dealt a blow to popular and philosophical convictions that there were passions and disorders typical of women, Jews, and "savages,"[59] and while others were typical of men. If we keep in mind that Idealist tradition either condemned strong passions as pathological or exalted them so far that they served heroic ends,[60] then we glimpse the heritage of "cultivated" German attitudes toward passions and pathology.[61] Like the unfolding psychiatry, popular philosophy believed that strong passions could be brought together with reason to produce virtue. If Romanticism cast doubt on the possibility of an enduring ethical Bildung, Freud's early work did likewise—as did his late work in the 1930s. The young Freud contributed to weakening the ethos of an irreflective sexual essentialism. After 1905, however, his contribution relapsed into the more familiar, sex-based distribution of passions and pathologies. A door stood open for a time, only to be closed again.

Notes

A version of this chapter appeared in *Traumatizing Theory: The Cultural Politics of Affect in and beyond Psychoanalysis*, ed. Karyn Ball (New York: Other Press, 2007), 1–40. I am very grateful to Professor Ball for permitting substantive portions of it to appear here.

1. I follow Didier Anzieu's distinction between trauma as a trigger and traumatism as the effect on and resultant condition, physical or psychological, in the victim. Although this difference is less frequently drawn in English than in French, it has value for us, because it underscores the radicality of Freud's etiology of hysteria. In Freud's medical "universe of discourse," where only the general concept of 'Trauma' existed, the idea that psychological events of a sexual or violent nature could leave a durable disorder behind them with somatic and psychological symptoms, was not considered when hysteria was in question. Tying hysteria to the effect called "trauma," with a resultant condition of "traumatism," allows Anzieu to distinguish more clearly between causes and effects. See Didier Anzieu, "Découverte par Freud du traumatisme sexuel précoce," *Journal de la Psychanalyse de l'enfant* 9 "Traumatismes," 1991. Hereafter cited in the text as *TSP*.

2. Christopher G. Goetz, Michel Bonduelle, and Toby Gelfand, *Charcot: Constructing Neurology* (New York: Oxford University Press, 1995), 187. Bourneville pointed out, on the basis of his many case studies that hysterics, in "their delirium . . . have remembrances [*réminiscences*] of long ago events in their lives, physical pains as well as psychological feelings [*des émotions morales*], events which have set off their attacks in the past." Freud and Breuer's remark that the hysteric suffers from "reminiscence" must be understood in this French psychiatric context. Hereafter cited in the text as *CCN*.

3. *CCN*, 261ff.

4. This is the famous "*globus hystericus*" or first theory of the origin of hysteria. See Sander Gilman, *The Jew's Body* (New York: Routledge, Chapman and Hall), ch. 3, "The Jewish Psyche: Freud, Dora and the Idea of the Hysteric," 60. Hereafter cited in the text as *JB*. Also see Sander Gilman, *Freud, Race, and Gender* (Princeton, NJ: Princeton University Press, 1993), 114ff.

5. See note 49 for Arthur Schnitzler's account of Freud's talk.

6. See Albert Lindemann, *Anti-Semitism before the Holocaust* (Essex, UK: Longman Press, 2000), 53. Lindemann writes, "From the mid-eighteenth century until the eve of the holocaust, the Jewish population increased faster than that of the non-Jewish population. . . . Their numerical rise was particularly striking: from 6,000 in 1860 to 175,000 in 1910, an increase of around thirty times within two generations. Budapest . . . experienced an even more precipitous increase in the same years, resulting in a Jewish percentage of 23 percent by 1914, compared to Vienna's 9 percent" (57). Hereafter cited in the text as *AH*.

7. Zeev Sternhell, *La Droite révolutionnaire, 1885–1914: Les Origines françaises du fascisme* (Paris: Seuil, 1978), 146ff. See chapter 3 whose conclusions—among which this one, that the French mistook nationality, culture, and social Darwinian "populations" for races arranged in a hierarchy—are borne out by Renan, who attempted to rectify them, see note 17. Hereafter cited in the text as *DR*.

8. *DR* 147. Sternhell adds, "The discovery of the unconscious at the end of the century contributes a complementary, even cardinal, dimension to the antirationalist and antidemocratic impetus. In this domain, the work of Gustave LeBon enjoyed a success almost unequalled to this day. His *Psychological Laws of the Evolution of Peoples*, first published in 1894, goes into its fourteenth edition in 1914, and his best known work, *Crowd Psychology*, which dates from 1895, goes into its 31st edition in 1925, and its 45th one in 1963. Translated into sixteen languages, LeBon's work . . . is one of the *greatest scientific successes of all times*" (*DR* 148; my translation).

9. See Laurence A. Rickels, *Nazi Psychoanalysis*, vol. 1: *Only Psychoanalysis Won the War* (Minneapolis: University of Minnesota Press, 2002). This is the first volume of a three-volume work.

10. Sander Gilman, "The Transmutation of the Rhetoric of Race into the Construction of Gender," *Freud, Race, and Gender* (Princeton, NJ: Princeton University Press, 1993), 36-48. Hereafter cited in the text as *FRG*.

11. See, for example, Catherine Hodeir, "Decentering the Gaze in Colonial Exhibitions," in *Images and Empires: Visuality in Colonial and Post-colonial Africa*, ed. Paul S. Landau and Deborah D. Kaspin (Berkeley and Los Angeles: University of California Press, 2002), 233–34, 240–41.

12. Daniel Boyarin, *Unheroic Conduct: The Rise of Heterosexuality and the Invention of the Jewish Man* (Berkeley and Los Angeles: University of California Press, 1997). Hereafter cited in the text as *UC*.

13. Gilman writes, "The very analysis of the nature of the Jewish body, *in the broader culture or within the culture of medicine*, has always been linked to establishing the difference (and dangerousness) of the Jew. This *scientific vision* of parallel and unequal 'races' is part of the polygenetic argument about the definition of 'race' within the *scientific culture of the eighteenth century.*" Many "Jewish" weaknesses were adapted from the pathogenic qualities of the female body which was "functionally pathologized—because it goes through phases of toxicity [*sic*] around menstruation, menopause, etc."

14. See Gilman's "The Jewish Genius: Freud and the Jewishness of the Creative," in *JB* 128–49. This different logic is tied to a binarism of creativity and neurosis. Freud struggled against both this and the identification of Jews with madness and sexual degeneracy. The Italian criminal anthropologist, Cesare Lombroso, contributed to the binarism with his work *Genius and Madness* (1864). See Gilman, *JB* 131. Note even there, in a work dating from eight years after Darwin's *Origin of Species* (1856), the social-Darwinian combination of "selections of species" and "race."

15. Michel Foucault, *The History of Sexuality: An Introduction* (New York: Vintage Books, 1990), see pt. 5, 133–59.

16. Georges Canguilhem, *The Normal and the Pathological*, trans. Carolyn Fawcett (Cambridge, MA: Zone Books, 1991).

17. Renan also remarked in his Sorbonne lecture "Qu'est-ce qu'une nation?" (March 11, 1882) that the great mistake of the present was to confound a race with a nation. The nation was a domain united under a dynasty, but he was skeptical about the notion of race, and blamed ethnography with its overuse; there were no pure European races. There were only races that became aristocratic through virtuous acts, and races that lost their virtue through similar means. Yet if race was not determinant for Europeans, Renan did insist, in an earlier work, that it had shaped the identity of Jews, see his *The Life of Jesus* (1869). So the question was really where a different race begins; who constitutes a separate "race." See Lindemann, *AH* 43, 129.

18. Gilman writes, "What *Sex and Character* did was to restate in a scientific, i.e., biological context, Arthur Schopenhauer's views on women and simply *extend the category of the feminine to the* Jew" (*JB* 133). For references to Schopenhauer, see Freud's "Formulations Regarding the Two Principles in Mental Functioning (note 3, 1911) and "Beyond the Pleasure Principle" (1920) in Freud, *The Standard Edition*, trans. and ed. James Strachey and Anna Freud (London: Hogarth Press, 1966). Hereafter cited in the text as *SE* with volume and page numbers.

19. Pierre Citti argues that in the French popular press, the theme of crisis overtook the rhetoric of degeneration by 1905, see *Contre la décadence:*

Histoire de l'imaginaire français dans le roman, 1890–1914 (Paris: Presses universitaires de France, 1987).

20. A case in point: the scandals about the alleged homosexuality of the chancellor and members of the second Kaiser Wilhelm's entourage in the Second Reich (1871–1918) reached the point of national hysteria by 1907. See "Iconography of a Scandal: Political Cartoons and the Eulenburg Affair in Wilhelmin Germany" in *Hidden from History: Reclaiming the Gay and Lesbian Past,* ed. Martin B. Duberman, Martha Vincinus and George Chauncey (New York: New American Library, 1989), esp. 234–35, 257–63.

21. Sigmund Freud, *SE* I (1886–99) "Pre-Psycho-Analytic Publications and Unpublished Drafts," 24ff. Notably his "Beobachtung einer hochgradigen hemianästhesie bei einem hysterischen Manne" (Observation of a Severe Case of Hemi-Anaesthesia in a Hysterical Man").

22. See Dianne F. Sadoff, *Sciences of the Flesh: Representing Body and Subject in Psychoanalysis* (Stanford, CA: Stanford University Press, 1998), 59–71. Hereafter cited in the text as *SF.*

23. The concept of 'irritation' (*Reiz, Reizbarkeit*—which also denotes "stimulation") dates from the emergence of physiology from anatomy. For a discussion of the notion of "irritation," see Didier Anzieu, *TSP* 22, 26.

24. Returning to hospital archives and original sources, Gladys Swain and Marcel Gauchet have argued, against Foucault, that the chief contribution of French psychiatry lay in the discernment of a "subject" of mania or madness, and that this "subject" could be solicited, under certain circumstances, by an experienced physician. See Gladys Swain, *Le sujet de la folie: Naissance de la psychiatrie*, (Paris: Calmann Lévy, 1987), 37ff.

It was Bourneville who, long before Freud's creation of his "talking cure," recorded patients' accounts, giving them the dignity of a narrative. Is it remarkable that his "scenes" closely resemble what Freud, later on, would hear? While no one yet spoke of traumatism, Bourneville recorded the words of a young girl, raped by her employer, who was institutionalized for hysterical attacks: "'You pig, you pig, I'll tell father . . . pig! You're so heavy; you're hurting me.'" See Marcel Gauchet and Gladys Swain, *Le vrai Charcot: Les Chemins imprévus de l'inconscient* (Paris: Calmann Lévy, 1997), 63–65. Hereafter cited in the text as *VC.* Here was Breuer and Freud's "reminiscence," dating this time from 1877.

Also see Anzieu, *TSP* 20: "The etiology of hysteria according to Charcot is double. The first cause is neuro biological: it is the nervous degenerescence proper to psychopathic families. . . . The secondary cause, which unleashes [hysteria], is tied to an event [*occasionnelle*]; its mechanism is essentially psychological." It was important to Charcot to maintain an inherited predisposition to hysteria, "degenerescence"—an obsession of European science in the last half of the nineteenth century.

25. Pierre Janet was Charcot's other illustrious student. Moreover, the question of how to understand this sexuality—as violation, as pleasure and violence—and its relationship to the will and to a multilayered memory was opened in this period. See Gauchet and Swain, *VC*, 66. Charcot's turn away from Bourneville appears to anticipate Freud's distancing from sexual traumatism following the demise of his friendship with Fliess.

26. Christopher G. Goetz, Michel Bonduelle et al., *Constructing Neurology* (New York: Oxford University Press, 1995), see chaps. 6 and 7.

27. Gauchet and Swain, "The Neurological Appropriation of Hysteria" in *VC* 49–96, esp. 49ff.

28. Charcot was quoted as saying privately that one should always expect "*la chose génitale*" to influence the etiology of female hysteria. Was this remark related to his colleague's transcripts and the work he pursued on demonic possession in art? In all likelihood. See Charcot and Richer, *Les Démoniaques dans l'Art: Suivi de la foi qui guérit* (Paris: Macula, 1984).

29. Charcot provided an impetus to Freud's interpretation of dreams by showing that hysterical "somnambulism" and dissociative personalities could be diagnosed as distinct from ordinary sleepwalking. The difference lay in the dream itself; hysterical somnambulism was found in cases where the dream proved irrecoverable by everyday consciousness. See *VC*, 176ff.

30. Freud distanced himself from conceptions in nineteenth-century neurophysiology, like degeneracy, but he remained a nineteenth-century thinker in his adoption, for psychosexual development, of the old recapitulation theory and a certain Lamarckianism. See Lucille Ritvo, *Darwin's Influence on Freud: A Tale of Two Sciences* (New Haven, CT: Yale University Press, 1990). Hereafter cited in the text as *DIF*.

31. The distinction "accident" versus "incident" is made by Anzieu, TSP 23. Anzieu writes, citing Freud: "From the notion of 'accident' we pass to that of an 'incident.' 'Any incident capable of provoking painful affects, fright, anxiety, shame, can act in the manner of a psychological shock,' at least in 'sensitive' subjects."

32. Gladys Swain and Marcel Gauchet, *Dialogue avec l'insensé: Essai d'histoire de la psychiatrie, précédé de À la recherche d'une autre histoire de la folie* (Paris: Gallimard, 1994), 230. Swain points out: "In fact, Charcot's ambition was initially . . . to do for hysteria what Jackson had begun to undertake for epilepsy. An initial problem was the differential diagnosis of the two diseases, with the vagary that recognition of a mixed entity like his hystero-epilepsy imposed. We have . . . from this inaugural parallel, a remainder . . . in the Freudian notion of *erogenous zone*. We know that the latter had for its origin the *hysterogenic zones* of which Charcot spoke. . . . Epileptogenic, hysterogenic, erogenous: in this verbal chain we have the . . . substance of the process that interests us."

33. Anzieu, *TSP* 23; the author is referring to Masud Khan's concept of "cumulative traumatism"; we already see this notion in its incunabula in Freud's November 1886 talk.

34. Freud, *SE* I, 24.

35. Freud added a precious insight into his own diverging views on hysteria in a brief essay entitled "Hysterie," first published in 1888 in the *Handwörterbuch der gesamten Medizin*, vol. 1, ed. A. Villaret (Stuttgart), 886–92, and reprinted in Freud, *SE* I, 39–57. There, he wrote, "The name 'hysteria' originates from the earliest times of medicine and is a precipitate of the prejudice, overcome only in our own days, which links neuroses with diseases of the female sexual apparatus" (41). There, he added, "Hysteria is based wholly and entirely on physiological modifications of the nervous system and its essence should be expressed in a formula which took account of the conditions of excitability in the different parts of the nervous system. A *physio-pathological formula of this kind has not yet, however, been discovered*; we must be content meanwhile to define the neurosis in a purely nosographical fashion by the totality of symptoms occurring in it . . . without any consideration of the closer connection between these phenomena" (41). Ironically, the idea that hysteria was based "entirely on physiological modifications of the nervous system" was later rejected by Freud, even before the "war neuroses" and "shell shock" demanded that the condition of hysteria, masculine hysteria, be reexamined.

36. Also see Gauchet and Swain, *VC* 65.

37. Freud, SE, "Introductory Lectures (1916–1917)"; "Lecture XXV Anxiety" cited by Lucille B. Ritvo, *DIF* 180.

38. Anzieu, citing Masson, lists three works: A. Tardieu's *Étude medico-légale sur les attentats aux mœurs* (1878), Paul Bernard, *Des attentats à la pudeur sur les petites filles* (1886), and Paul Brouardel, *Les attentats aux mœurs* (1909). Anzieu, *TSP* 24.

39. Of this censoring function Freud wrote in 1912, "[We] learn that the unconscious idea is excluded from consciousness *by living forces,* [these] oppose themselves to its reception, while they do not object to other ideas, the preconscious ones." See his "The Unconscious in Psycho-Analysis," in *Sigmund Freud Collected Papers*, vol. 4 (London: Hogarth Press, 1949), 27.

40. See Freud's observations for the 5th International Psychoanalytical Congress in 1918 "Introduction to the Psychoanalysis of War Neuroses" in Freud, *Gesammelte Werke*, 1917–20, 323–24. Also, his 1920 remarks in *Beyond the Pleasure Principle*, "In the case of war neuroses, the fact that the same symptoms sometimes came about without the intervention of any gross mechanical force seemed . . . bewildering. In the . . . ordinary traumatic neuroses, two characteristics emergence prominently: first, that the chief weight in their causation seems to rest upon the factor of surprise, of fright . . . and secondly, that a wound . . . inflicted simultaneously works as

a rule *against* the development of a neurosis." In Peter Gay, ed., *The Freud Reader* (New York: Norton, 1989), 598.

In both cases, one of the concerns is to hold war neuroses separate from "ordinary traumatic neuroses," and to maintain both of these in contraposition to transference neurosis. The term "hysteria" does not appear, but we can assume it belongs to the transference neuroses.

41. Juliet Mitchell explores the mimetic dimension of hysteria, which she perceives in the Freud-Fliess relationship. Her argument concerns the omnipresence, today as in 1900, of hysteria, in women and in men, where it can take the form of "Don Juanism." See Mitchell, *Mad Men and Medusas: Reclaiming Hysteria* (New York: Basic Books, 2000), 251. Hereafter cited in the text as *MM*.

42. It is a bodily unconscious both through its physical symptoms and given Freud's insistence on the relationship between external events and internal excitations that must be discharged through speech or the muscles.

43. Freud wrote in the Spencerian-Darwinian language that would long be his own, that the overflow of excitations within the neuropsychic system constitutes the "prototype . . . of *psychical repression* (*SE* 5, 600), cited in Ritvo, *DIF* 187. He added, in his "History of the Psycho-Analytic Movement" that "the theory of repression . . . is the corner stone on which the whole structure of psycho-analysis rests" (*SE* 14, 16).

44. That is, investments of psychic energy directed toward self in the case of a conflict between desires or in that of a nonrepressed, physical wound, see note 24.

45. However, against Foucault's arguments in *The History of Madness* (1961), Swain claims that the psychiatric innovation entailed a revolution in identities whose primary upshot was a profoundly different attempt to include the "mad" within society at large. The psychiatric revolution, however, had for its effect, transposing an external monstrosity into a generalized immanence; under the right circumstances, one could indeed speak with the mad, and so the monstrous either did not really exist, or resided within each of us. See her *Dialogue avec l'insensé*, 114, 118.

46. Freud, "A Difficulty in the Path of Psycho-Analysis" (1917), in which he identifies psychoanalysis as the third "death blow" to man's narcissism. Cited in Ritvo, *DIF* 22ff.

47. Freud "identified" with a number of heroic figures, from Moses, carrier of the Law, to Hannibal and Oedipus. Tracy B. Strong discusses this in "Psycho-analysis as a Vocation: Freud, Politics, the Heroic," *Political Theory* 12, no. 1 (February 1984): 51–79.

48. Freud, *SE* 20, 10. Cited by Lucille Ritvo, *DIF*, chap. 12, 170–71. Following his presentation on male hysteria, Freud was excluded from Meynert's laboratory of cerebral anatomy. Freud, *SE* 20, 15; cited in Ritvo, *DIF* 171.

49. Ritvo, *DIF* 170. In another account of Freud's October 1886 lecture, this time by playwright Arthur Schnitzler, the question does not turn on the impossibility of male hysteria, and Meynert and others are even recorded as saying that there were documented cases in Viennese hospitals and German medical journals. See Gilman, *FRG* 116.

50. Charcot and his associate Bourneville (see notes 24 and 25), first tied sexual dysfunction and even the experience of sexual violence explicitly to the etiology of nervous diseases. The question that remained was how the neurological (modifications to the spinal nerves mainly) becomes the cerebral, and "where" one situates the "appareil psychique." See note 35.

51. Mitchell, *MM* 74. Mitchell is referring to the claim that, after his break with Wilhelm Fliess and the publication of the aborted Dora case (1905), Freud's concern with hysteria waned.

52. Judith Herman conversation, Radcliffe Institute for Advanced Study, February 21, 2002.

53. For an extensive discussion of the visual construction of the African in the nineteenth century, see Paul S. Landau and Deborah D. Kaspin, eds., *Images and Empires: Visuality in Colonial and Postcolonial Africa* (Los Angeles: University of California Press, 2002). For historical analysis of these stereotypes, see Gustav Jahoda, *Images of Savages: Ancient Roots of Modern Prejudice in Western Culture* (New York: Routledge, 1999).

54. For a historically sensitive account of the "traumatic neuroses of war," see Judith Lewis Herman, *Trauma and Recovery: The Aftermath of Violence, from Domestic Abuse to Political Terror* (New York: HarperCollins–Basic Books, 1992), 20ff. Hereafter cited in the text as *TR*.

55. Weininger's *Sex and Character* sold off the shelves on the news of his suicide in the fall of 1903. His defenders included Wittgenstein. See Jacques Le Rider, *Le Cas Otto Weininger : Racines de l'antiféminisme et de l'antisémitisme.* (Paris: Presses Universitaires de France, 1982), 144.

56. I am speaking of Leibniz and Spinoza, and their less illustrious nineteenth-century philosophical heritage in figures such as Fechner and Herbart.

57. When Freud treated the Russian aristocrat he called the "Wolf Man" in 1918, it was for obsessive compulsion. Juliet Mitchell points out that "Freud's comment on the Wolf Man [and his hysterical fantasy about bearing an infant in his bowel] is not expanded into further thoughts on male hysteria. . . . However, just as important is his assertion from observing the Wolf Man that, beneath all other neurotic disturbances in man or woman, there is a layer of hysteria." See Mitchell, *MM* 71.

58. Mitchell, *MM* 64–65. Drawing on the work of Anzieu and Eva Rosenblum, Mitchell points out that "E" was a patient of Freud's for five years. She adds that "'E' and Freud also had a number of symptoms in common and shared certain aspects of their reconstructed infantile histories. 'E"s fits of profuse sweating, tendency to uncontrollable blushing and dread of

going to the theatre were traced back to his fantasy that he would 'deflower' every woman he set eyes on. Freud, too, had a crucial 'deflowering' incident in his history. 'E' had failed botany at the university and Freud commented, 'now he carries on with it as a deflorator.' Freud remembered how he too was a deflorator. When he was a small child, he had snatched and destroyed his niece Pauline's yellow flowers. This incident became a crucial part of the ground plan of Freud's own later fantasies of defloration."

59. The literature on the emotiveness and "infantilism" of non-Western peoples is vast. See note 53 for two sources.

60. Kant's lectures in *Anthropology* compared emotions and passions to bouts of influenza, which one had to weather or sleep off. Hegel praised certain passions in his *Philosophy of Mind* (*Encyclopedia*).

61. To an extent, these were also the attitudes of the French. However, from the mid-eighteenth through the nineteenth centuries, materialism in France—from de la Mettrie to Pinel and Charcot—inflected French attitudes either toward physiological explanations of the passions or toward a different romanticism in their regard.

Part 5

Afterword

12

Terror's Wake

Trauma and Its Subjects

Michael Lambek

Oedipus: *O and again*
That piercing pain,
Torture in the flesh and in the soul's dark memory.

Chorus: *Twice-tormented; in the spirit, as in the flesh.*
Creon: *There is a measure in all things.*

—Sophocles, *King Oedipus*

I write in the aftermath of the tsunami in the Indian Ocean: an estimated one hundred fifty thousand dead (and rising) and countless more who experienced the terrifying surge of water and are now bereaved and bereft. Is it right or useful to call the experience of the survivors "traumatic" or refer to it as "a trauma"? Would it be wise or presumptuous to medicalize their suffering and identify it as post-traumatic stress disorder (PTSD); callous or practical to ignore their emotional and existential suffering and focus only on meeting material needs? What edge does trauma give over other models of understanding and intervention? What risks might it entail? Whose meaning would it affirm? What does it add to our understanding of human subjectivity?

We wish for a world of no suffering, but our biology, psychology, history, and society all mitigate against that. Across human history women have given birth in pain. A percentage of women die in childbirth or suffer damage.

Human infants suckling from live mothers found their subjectivity in eventual loss of the breast. The family is a nexus of ambivalent emotions; some children everywhere grow up angry, conflicted, or depressed. Some young men become abusive to women. Socially, many things seem to have grown worse. Through much of history, humans ate relatively well and, apart from seasonal fluctuations in some climatic regions, did not experience great hunger. Anthropologists have described societies based on a foraging mode of subsistence ("hunters and gatherers") as "the original affluent society" because their means were sufficient to meet their ends.[1] Chronic malnutrition is largely a product of European expansion and has risen to its greatest proportions in the age of industrial capitalism. Poverty and wealth evolve together, through forms of "structural violence" characterized by appropriation, unemployment, and exploitation. Warfare too was never greater than over the past century, its brutality never more horrifying, though this is surely not a matter of competition. Twentieth-century violence, whether of the gas chamber, the nuclear bomb, the torture cell, the sex trade, or the pervasive spread of land mines, has been calculated and systematic. Social suffering has become endemic;[2] and witnessing or learning about suffering has too. The current North American epidemic of depression is likely linked to the marketing of medications ostensibly capable of treating it,[3] but presumably as well to many other social factors.

We need to know the world in order to change it. We can work concretely on such fundamental issues as documenting the production and traffic of land mines; capitalist schemes that bring widespread environmental damage and economic misery; social conditions and traditions that denigrate, isolate, and disempower women and children. We also learn to address and redeem personal suffering; and through religion and healing to accept what we cannot change, and gain, as the saying goes, the wisdom to know the difference; to sublimate suffering in art or eschatology. But can we "know" suffering in the same way that we come to know the facts of nature or the economy?

How we know the world affects how we act to change it. Since the nineteenth century we have developed a secular and scientific way to address suffering and the anxiety it produces, namely, to objectify it. We divide, name, and classify suffering; diagnose, measure, model, and count it, all the better to tame it. By 1924, Thomas Mann could gently ironize this impulse in his depiction of Herr Settembrini in *The Magic Mountain*, who from the (literally) rarified air of the tuberculosis sanatorium in Davos contributes to a vast encyclopedia of suffering characteristic of the "flatlands" below.[4] The inhabitants of the sanatorium go about their daily regimen of walks and meals in the face of certain or uncertain degeneration of their lungs as the world below prepares for the next war. Mann's characters articulate Marxist and Freudian insights in their debates with one another even as they lack insight into their personal conduct.

Objectifications, such as Frankensteinian monsters, sometimes take on a life of their own and may even contribute to the very effects they were designed to suppress. Distinguishing the consequences of objectification has been the object of thinkers such as Michel Foucault and Ian Hacking.[5] They look at the power of experts to produce authoritative discourse, of discourse to produce the very objects of which it speaks, and of discursive objects and practices to impose themselves, to subvert and transform human subjectivity. They look also at how the ways we think we know the truth of things are themselves highly contingent, how forms of "redescription" can change the very nature of certain human experiences and conditions. Hacking asks how, with the language of subjectivity we have for ourselves today (a language that includes such words or concepts as "trauma"), we can truly describe or know the subjective worlds of others, living at other times, in other places, construing themselves by means of different terms and different practices. Is it valid to redescribe the shell shock of World War I soldiers as posttraumatic stress? What is gained and what is lost?

We don't need complicated language or particularly subtle thinking to see objectification in operation. An undergraduate student returned some years ago from an internship in international development in which she was placed with mental health care workers in northern Uganda. These workers were busily diagnosing and treating PTSD among children who had been kidnapped from their homes and forced by rebel soldiers to commit atrocities. What the aid workers did not fully grasp (or want to grasp), asserted the student, was the condition of chronic uncertainty that prevailed in northern Uganda. People who had fled their abductors could be recaptured, violence could break out from either side, and people were moved around as pawns of more powerful forces. The treatment of PTSD is predicated on a return to "normal" society, a return from the front, the ability to contain violence. It also assumes a distinction between individuals and their social context. Of what use are these categories or interventions in the midst of such a protean conflict, such deep instability, such pressing economic need? Further, in concealing insecurity, do they not collude in its effects and possibly even with the structural forces that lie at its origins? Just whose anxiety do they address?

Some would argue that Foucauldian analyses themselves depoliticize and close down forms of intervention, others that they enhance our understanding and our politics, that a sort of Foucauldian revolution has enabled the progressive resubjectification and agency of those abjected by powerful and pervasive discursive regimes, including ostensibly well-meaning regimes of development or therapy. I do not myself wish to arbitrate among different forms of liberal, Marxist, feminist, or poststructuralist analysis of politics, merely to signal that they may differ precisely in the means by which they conceptualize suffering. The Foucauldian point is that authoritative conceptualizations and practices—of any form—have particular effects.

My own perspective is that of an anthropologist interested in dialogue with philosophy and having the sense that our respective fields share more than we often recognize. We are both interested in the human condition and the ways in which it is mediated by the production and limits of meaning. I have worked in societies that conceptualize suffering quite differently from the understandings prevalent in contemporary psychiatry and this has led me to want to question certain Euro-American ideas and practices. However, I find it much more difficult to achieve the same measure of critical distance and hermeneutic proximity "at home" as I strive for in understanding the practice and performance of spirit mediumship among Malagasy speakers in the western Indian Ocean.[6] Spirit possession as a cultural system draws on the human capacity for dissociation in creative and ethical ways that, on the one hand, challenge me to find the language within the Western tradition to comprehend them while, on the other hand, provoke skepticism with respect to some of the cherished ideas of that tradition.

Principled skepticism is something philosophy and anthropology ought to share but our questions often arise on different grounds. In what follows, I attempt to do justice to some of the ideas raised by the authors of this volume, adding questions that an anthropologist might ask of trauma.

The first question this set of lively, smart, and diverse chapters invites us to address is why trauma should be of interest to philosophers and others not directly involved in therapeutic intervention. Is there any advance to be had on traditional philosophical problems from the recent attention in popular and medical circles to trauma? Trauma is a puzzle for theories of the self-conscious rational subject capable of transparent, consistent, and objective self-representation. Conversely, it may offer a kind of tool for those who wish to challenge such theories or champion the alternatives. The chapters in this book cumulatively teach us that trauma interests philosophers insofar as it provides a window—or perhaps a cellar door—into human physicality or corporeality; into aspects of mind and body that are evolutionarily deeper, ostensibly, than symbolic activity and reflective consciousness—deeper, therefore, than self-understanding, or at least external to it. Trauma offers a class of experiences that are not integrated within the ego or language. For those who make their way with words, symbols, and self-understanding, this is novel, dangerous, and hence fascinating terrain.

However, the authors do not agree on the lay of the land or how best to navigate it. Charles Scott, in whose chapter the connections between mind, body, and self are elegantly made, describes how the "I" can be located at a distance from the trauma and can appear to observe it. Idit

Dobbs-Weinstein makes the distance a much more radical one: trauma is not available to conscious experience precisely because it overwhelms the very possibility for experience. For Dobbs-Weinstein, in trauma the "I" does not merely retreat to safe ground, it disappears. These authors illustrate, respectively, the antimimetic and mimetic approaches that Ruth Leys argues have structured the history of the concept of trauma from its outset. Their contributions also illustrate what Leys has observed as the inevitable collapse of each of these theories into the other.[7]

A key feature for Scott is indifference. He unpacks several senses of the term. First is the indifference enabled by the distance of presentation itself, "the availability of trauma for nontraumatic, perceptive experience" in various forms of media. Second is the indifference characteristic of a kind of dissociation experienced by people who report watching themselves drown; situations in which people who know themselves to be in danger experience a kind of helpless detachment. Third, and most critical, is the indifference that creeps from memory into daily experience, an "indifference to place and time . . . that makes normal daily living impossible." Scott says, "Something without value one way or the other appears—something without character or personality, without clarity of interest, intelligence, or choice. I am noting the indifference of an instinct severed from the partiality and interests of complex human awareness and values." For Scott, this is because "trauma happens as a somatic disturbance." It registers as a "prereflective memory trace" without spatial and temporal context and hence the trauma is present again, "whenever something triggers this timeless, placeless memory." This occurs "in blind inappropriateness for given circumstances and in a destructive noncoordination with the abilities of social consciousness and self-direction." Scott thereby grounds or contextualizes his acute phenomenology of indifference in brain biology. Interestingly, however, his conclusion is not deterministic; he argues for differential responses and the possibility of resilience, for diminishing the effects of trauma and making good our losses.

Dobbs-Weinstein takes a different tone and locates her argument in a different intellectual context. She writes, after Adorno, about experience and its barest possibility after Auschwitz. She argues against the abstractions of Kantian philosophy and its advocation of hope as forms of escape from "unbearable, unknowable, singular experience." She expounds on "the complicity, perhaps identity, between religion/metaphysics and the redemptive history that not only justifies suffering but, more important, immunizes against . . . its experience." Dobbs-Weinstein offers a sharp critique of the subject, yet one of the interesting lessons of her argument is that one must address the impact of injury not only on victims but on perpetrators and witnesses. A corollary is that we must not produce self-justifying analyses in which we implicitly identify with victims; we must explore also the ways in which we function ourselves as perpetrators and as witnesses.

Scott and Dobbs-Weinstein reach different conclusions. Dobbs-Weinstein celebrates "the poet's courage to resist the temptation of making the events intelligible." Scott asks rhetorically, "Would we want traumas portrayed traumatically? Would we want nonvirtual trauma, real trauma in the presentation?" The value of indifference is "best not moralized and criticized while we benefit from it." I share Dobbs-Weinstein's concern that the concept of 'trauma' can serve an immunological and normalizing function. But the task to portray trauma directly (let us call this to "present" trauma), like Dobbs-Weinstein's poet, rather than to portray it indirectly (let us call this to "represent" trauma), like Scott's virtual trauma in the popular media, is not one all of us can follow; it is not the path that philosophy itself follows.[8] And it is only under some form of "religion/metaphysics" that reveals or elaborates the value of fully confronting suffering that it would be deemed necessary that all of us present or receive such traumatic portrayals. Complex issues of aesthetics, ethics, and politics are entailed in this debate that have relevance both for public culture and for articulating the means and goals of individual therapy.

Kristen Brown Golden draws another philosopher to the table with her invocation of Merleau-Ponty's transcendence of Cartesian dualism. Rather than use Merleau-Ponty to critique current concepts of trauma, Golden intriguingly uses research on psychological trauma, specifically spinal cord injury, as evidence to support or defend "Merleau-Ponty's view of humans and nature, bodies and psyches as interpenetrating."[9] Hence she advocates a shift from conceiving of "self-contained faculties with defined functions" to models of "interpenetrating coresponses to felt bodily needs." But if human bodies are self-regulating systems, as they undoubtedly are, how far can such a model incorporate conflict? Certainly, although Freud attempted such a model, it was insufficient to account for mind's disruptive qualities.[10]

Golden's powerful argument that communication "has roots in the body" can be complemented by Geertz's argument (described later) that the human body and mind have also been shaped by our species' use and dependence on language. Moreover, whether, as Golden argues, the main function of communication is to show oneself or whether display is balanced with other concerns, such as concealment, are open questions. One strategy for exploring them would be to take a Peircean line in distinguishing indexical from symbolic signs and to link these differentially to such questions of function and intention (or presentation and representation).[11] Golden's chapter also implicitly raises intriguing questions about metaphor, specifically, how deeply into the body we want to push metaphors for thought, intention, and practice. Conversely, how literally ought we to take the idea of corporeity as communication? Is metaphor indispensable here, and if so, what does this say about the mind-body divide? Are specific metaphors illuminating or harmful?[12]

No doubt Merleau-Ponty ought to be given greater exposition in discussions of habitus or embodied memory that is exhibited in such things as bike riding, posture, and even composure, as well as the ways people dwell unselfconsciously in the kinds of places they make for themselves. Trauma would appear to threaten such primary intentional stances and trauma therapists might be advised to investigate blockages in embodied practices no less than in psychological memory and mental imagery.[13]

Any mention of dwelling evokes Heidegger. For Eric Nelson it is the question precisely of the "unhomely" (*unheimlich*) in Heidegger that must be called to attention for thinking about trauma. Nelson offers a careful reading of Heidegger's *Introduction to Metaphysics* alternative to that of Levinas. Rather than valorizing violence, suggests Nelson, Heidegger is emphasizing "the possibilities for responding" to the violence intrinsic to human being. This violence is not at base a conflict between humans but "a question of being itself and of human existence." It is not a willful or social Darwinian struggle for existence, but "an originary strife . . . that is intended to contest the reification of identity." Nelson argues that the responsiveness Heidegger advocates is precisely away from self-assertion and struggle, toward release and composure (*Gelassenheit*).

That pain and even violence lie at the origin of being and difference and require some kind of authentic response is a profound thought—but it continues to lie at some distance from most discussions of trauma, which understand by that term the intervention of specific acts and contingent events and which thereby distinguish the victims of trauma from other humans (or human being), including perpetrators and witnesses. A lesson for trauma theory is perhaps that it ought to question its own tendencies toward the violence of reification.

It is evident from each of these chapters that an obstacle or at least puzzle for understanding human suffering is the dualistic mind-body lens we have inherited from Descartes (and with much deeper roots, as my epigraph from Sophocles indicates).

People everywhere, but some places more than others, somatize suffering. We say then that psychological anguish is expressed in bodily symptoms—hysteria, neurasthenia, headaches, and bowel disorders. Some go much further, linking stress to heart disease or lowered immune system responses. But conversely, of course, people subjectify or psychologize events in the physical world—grieving, angering, worrying, harboring grudges; even repressing, dividing, depersonalizing. It is difficult to say that one of these processes is secondary to the other. Certainly, if we were to transcend the mind-body opposition, "somatization" would not be understood as a secondary process, a transformation of what is first psychological. Nor would traumatism be understood as a matter of transforming embodied experience into mind. Nowhere, it seems, is this nexus tighter than in experiences of sexuality and violence that impinge simultaneously on our bodies, our fantasies, and our sense of trust in human relations.

The question is whether trauma theory provides the opportunity for a revolt against the precedence of "mind" in current thought and hence a reversal that places the weight on "body"; whether it is an attempt to restore balance between mind and body; or whether it offers a true transcendence of the opposition. Those who wish to emphasize embodiment need to think very carefully, as do Kristen Golden and Charles Scott, through the consequences of allying their models with those of the neurosciences.

A monistic approach that emphasizes commensurability and connection risks a reduction to physicalism. I would take a different direction. Dualism is surely not wrong in an absolute sense or it would be neither so durable nor so widespread (in various forms) cross-culturally. It is, to some degree, indicative of the complex human and natural reality underlying it, a reality that is not directly graspable by a unitary commonsense model or single perspective. What is "wrong" about Cartesian dualism is exactly its literalness, the way it objectifies mind and body, reifies the distinctions between them, and tries to police the borders. I understand the human condition as simultaneously one of both mind and body, understood as two incommensurate perspectives, rather than a consistent unity or an explicit and clearly demarcated boundary. Mind and body invite for me not a literal division of territory but a recognition of irony, that is, a fundamental uncertainty concerning the ground of human existence and hence of knowing oneself.[14]

I have argued elsewhere that mind and body are incommensurate precisely (and perhaps paradoxically) because if we start from the perspective of body they appear commensurable and unified (as evident in recent work in neurophysiology or neuropharmacology), but if we start from the perspective of mind (as evident in structuralism's binary categories of thought or in a variety of local/cultural distinctions between body and mind, spirit, or soul) they are not.[15] To accept incommensurability is to accept that we may not be able to know with any certainty whether a given state is best described as mind or body. That is to acknowledge the irony of the human condition; that to "know oneself" must be balanced with the wisdom of not knowing, of accepting what one cannot fully know or accepting that some questions about oneself cannot be answered with certainty. Put in another language, it is to balance "certainty" (*certum*) with "truth" (*verum*).[16]

A further advantage of acknowledging incommensurability is that, if applied dialectically, it offers a means to conceptualize change. If recovery is too hopeful a word, it enables at least shifts in perspective so that subjects do not remain overwhelmed by events but are, in fact, able to purchase some distance from them. We simultaneously live experience and reflect on it; that is what distinguishes human consciousness from the animal. Our lives are constituted in dialectical moves between lived experience and its objectification in language and narrative rather than in fixing the relations between them.[17] One might, then, describe trauma sufferers as caught in literalism; it is as though the effect of trauma is to incapacitate our sense of irony.[18]

Anthropologists understand culture to shape and mediate human existence. In some of his early essays Clifford Geertz has written profoundly about the intrinsic connections between human nature and culture. The human brain and body have evolved in tandem with an increasing dependence of the organism on culture, hence the brain is shaped by culture and language no less than culture and language are shaped by the brain. Therefore, asserts Geertz, it is a profound mistake to assume that we can simply peel back culture and discover our intrinsic "nature" preserved intact beneath it. Given the adaptation of the human brain to culture and language, Geertz argues, human beings could not even think or feel without culture.[19] In a parallel argument, Marshall Sahlins describes culture as the human form of mediation; our very perceptions are culturally shaped, such that, for example, "appetite" mediates or even supercedes biological "hunger" and hungry Anglo-Americans are nauseated by dog meat much as (but for different reasons) Hindus reject beef.[20] The culture concept thereby has significant implications for thinking through the relationship of mind and body.

The concept of culture appears to approach its limits when describing disorder or disruption, when experience overwhelms existing cultural codes and frames (and indeed, the notion of cultures as holistic, orderly, bounded, and individuated units is now in considerable disarray).[21] And yet there are extensive literatures describing the interpretation and transcending of both social anomie and personal misfortune by means of the cultural imagination, carried out in such practices as song, poetry, ritual, and religious revival.[22] Culture is protean, resilient, and renewable; it abhors an existential vacuum.

This suggests that the results of (some) traumatic experience need not be as shattering, as permanent, or as amplifying as they are described by Sandra Bloom. At the same time, however, some form of social order is necessary if cultural production is to flourish. The sense of personal well-being or resilience is related to the broader social context and, as Bloom notes, the kinds of security, predictability, hope, and confidence it enables.

All this is nicely illustrated in the chapter by Galaty, Stocker, and Watkinson. Freud might have appreciated the metaphoric qualities of the bunkers that cover the Albanian landscape like cysts. Most obviously a sign of vigilance and even a "siege mentality," of containment and entrapment, they form an interesting image for the whole subject of trauma. As the authors note, they not only protected against external foes, but simultaneously established a system of internal intimidation. However, Albanians have liberated themselves from this internal oppression and turned their bunkers into sites and signifiers of lovemaking, tourist art, and parody. Similarly, the practice of piecing together new buildings from the ruins of much older ones that the archaeologists describe forms an engaging metaphor for "healthy" memory, in which elements of the past change their meaning as they are applied to new contexts.

Several authors in this collection define trauma in terms of a temporary but absolute loss of meaning (culture) and hence, as Geertz suggests, of the very capacity to think and feel. They provide sophisticated ways to articulate the inarticulate, the absence of meaning in human experience. Victims—and perhaps even perpetrators and witnesses—may be conscious in a material, neurophysiological sense, but not self-conscious in the reflective, fully human, sense of intentional, meaningful engagement with the world. In psychological terms, as Sara Beardsworth puts it, "A trauma is an event that overwhelms the ego's defenses." Therefore, as she says, it cannot appear within the terms of a subject and object distinction, and is not present to consciousness.

However, this experience—or nonexperience—is necessarily situated with respect to the world of meaning, including the ego, that surrounds it and which it disrupts. Distress occurs when this resituating is not one of integration. Beardsworth again: "With Freud, traumatic suffering is caught in a present that is nonfutural because the sufferer cannot take up a relation to what has occurred. Traumatic suffering is structured by the persistence and dominion of a past within the present, turning it into an endless present."

In going beyond meaningful, reflective consciousness and in arguing for the effects of some kind of unconscious trace or "memory," each of the authors must in some fashion acknowledge Freud. Freud is among the earliest and certainly the most persistent modern writers to explore the terrain of the unconscious, to take seriously the fact that reflective consciousness is radically insufficient to explain human subjectivity, desire, or symptom; in sum, to show that our conscious rationalizations consistently fall short in explaining our behavior and that our motivations are not transparent to ourselves. Culture may be orderly and meaningful, but human beings are conflicted, grappling with public worlds of meaning in the face of forces that remain fundamentally disruptive, disorderly, and inaccessible to reflective consciousness. And yet, theorists of trauma stand in an ambiguous and conflicted relation to Freud.

This is, very simply, because Freud found most fascinating the wounds that are self-inflicted or that arise internally or ontogenetically, within the everyday relationships between infants and their parents. He saw this nexus as the source of human happiness, of the capacity for trust and contentment, but also as the source for unhappiness and the seedbed for neurotic misery. He emphasized the prevalence and perversity of human fantasy and noted again and again how fantasy becomes lived reality in human experience. In Freud happiness is never the whole picture.[23] Whatever their views on happiness or fantasy—and these remain largely underarticulated—theorists of trauma emphasize wounds that are other-inflicted. Sometimes they come from fully external, neutral forces, such as the tsunami, but most saliently, they come from human agents, in acts of deliberate violence and violation. In sum, whereas the Freudian subject is characterized by ambivalence and Freudian theory is ambiguous about distinguishing internal from external sources and forces, the subject in trauma theory is characterized by an original (possibly undertheorized) wholeness that is fractured through intervention by explicitly external agents and events.

The distinction I am making here between Freudian theory—based on internal conflict and injury—and trauma theory—based on responses to external events—is too simple because Freud also theorized about the internal effects of external events; indeed, in looking over his corpus chronologically, one can ascertain more than one point at which he proposed a theory of trauma. However, the discursive landscape has proved a good deal murkier than a simple textual analysis of Freud could address. This is because Freud himself (and the psychoanalytic movement that he founded) have come to carry tremendous symbolic import and there is much struggle over what is invoked in their name. A number of theorists and therapists who emphasize the empirical prevalence of sexual abuse and who see abuse at the source of a wide range of symptoms accuse Freud of having abandoned the evidence of trauma; these "trauma theorists" thus pose their model in direct opposition to Freud's. For them, therapy is a site less for exploring the patient's conflicts or fantasies than for uncovering the historical truth. In the influential work of Judith Herman, the obligation of the therapist is to believe her patients and, when they are silent, to help them uncover original traumatic events and confront the perpetrators. This kind of therapy helped produce "recovered memories," which in turn rapidly gave rise to debate about their plausibility. But whereas the supporters of recovered memory attacked Freud, so too did some of the skeptics, who blamed Freud for providing the very models of repression and therapy that made the work of trauma advocates possible.[24] Poor Freud got it from both ends and his own ideas were often reduced to stereotypes in which it was their emotional import rather than their content or progression that was of significance.[25]

Sara Beardsworth and Gregg Horowitz each take this debate away from polemics and back to a balanced and closely reasoned interrogation of specific areas of Freud's thought and of psychoanalysis. On the one hand, they acknowledge and explore the ways in which a concept of trauma is central to Freud; on the other, they are concerned with the ways the kind of thinking about trauma I have mentioned might infiltrate psychoanalytic or psychotherapeutic thinking and practice. In particular, they want to restore the Freudian concept of object loss and the genealogy of subjectivity. They do not wish to see the concept of object loss reduced to that of trauma.[26]

Beardsworth places the source of confusion partially in Freud himself. In her bracing reading of *Moses and Monotheism*, historical trauma is transformed subjectively into personal loss and yet this loss is itself "lost," that is, forgotten. Loss is "rediscovered" in Kristeva's depiction of the preverbal—and preoedipal—forms of separation of the infant from the mother. As Beardsworth says, "The primal loss of the other is not an overwhelming exposure to exteriority marked by a lack of affect. It is not a trauma." But in Kristeva's diagnosis, the forms of corporeal and affective responsiveness characteristic of the loss-inflected maternal bond are neglected in the secular, rationalized, and paternal/oedipal world of modernity. Hence a possible implication of this chapter is that the dissatisfaction widespread in North American society and picked up now (or redescribed) by trauma discourse might better be understood as a symptom of the failure to activate and recognize the affective dimension of subjectivity founded in primary loss.

Beardsworth herself goes a step further, arguing that the very transition to modernity may have the structure of a trauma and hence that our attachment to trauma theory at the expense of recognizing loss may be itself a symptom of our historical and collective "posttraumatic" condition. Evaluation of such a stimulating hypothesis requires closer attention than I can give here both to the effect of positing a sharp distinction between "modernity" and its other (or others?) and to the legitimacy of "social Freudianism" (on the analogy to social Darwinism), that is, of theory that poses analogical or causal relationships between individual and collective processes, events, and conditions. But her argument grants a surprising weight to trauma itself.

Horowitz makes the case much more elegantly and cogently than I have been able to for the limits of an objectivizing science of mind or indeed for "a complete, consistent, and singular metapsychology." He argues intriguingly that Freud "realized the aim of psychoanalysis had to be not merely to achieve Socratic self-knowledge but rather to strengthen the psyche in its combat with an authority [paternal] that can sometimes appear in the form of the demand to know yourself." Thus the point of psychoanalytic therapy is not to reveal hidden knowledge—which for Horowitz is actually patent—but to understand why we continue to discredit, and suffer, it.

Horowitz argues that psychoanalysis sees this "ability of the present to hold on to the undead past" as "the mark of normal suffering" and that "[n]ot transcending or redeeming loss, not ceasing to suffer, but sustaining loss, continuing to suffer creatively, is the mark of healthy individuation." But this is precisely what the trauma victim cannot do. Instead, as he quotes Stephen Prior, trauma is a matter of "relentless reliving" of an all too comprehensible and real event. Horowitz makes a clear distinction between (healthy) loss—of which psychoanalytic theory provides an account—and radically disabling trauma, portrayed as "the incessance of injury." Horowitz rejects the identification, by theorists, of loss with trauma—not only because it obscures the nature of loss but also because it denies "what in trauma is traumatic."

I like Horowitz's distinction but I also think that it does not apply to much of what now goes under the label of trauma. Horowitz offers a powerful portrait of the traumatic condition, epitomized for him in childhood sexual abuse. But such "soul murder" is hardly the same condition that might be experienced by someone surviving a traffic accident or even warfare.[27] The inability to form new object relations is not the same condition as being troubled by intrusive images or nightmares. Furthermore, we should not underestimate the difficulties of discerning trauma or distinguishing it from loss in much actual experience. One of the most poignant images from the wake of the tsunami is of mothers overcome with pain and grief that their children slipped from their arms. What these women show is not a damaged ego. There may have been terror, and there may be the relentless reliving characteristic of trauma, but there is also guilt and loss. Over time, with the work of mourning rather than with any specific "trauma therapy," their "relentless reliving" may be more or less well transformed into "normal suffering."[28] Refugees, too, may have experienced terrifying events, but they also experience profound loss, where what has disappeared includes not just the meaning or memory of a horrific event but the context for interpretation, the very world in which certain events could be given meaning.[29]

"Trauma," suggests Horowitz, "is contagious." Assuming he means here the discourse of trauma (and hence the diagnosis), I could not agree more. But while Beardsworth and Horowitz question the elision of trauma with loss that may occur in psychoanalytic theory, the question remains, What happens outside it? Is there a therapeutic practice specific to trauma? Over what range of cases ought it to apply?

Bettina Bergo's discussion of hysteria offers a kind of mediation of this debate over the scope of trauma insofar as hysteria itself mediates between trauma and loss. Hysteria is critical because it forms a historical predecessor to post-traumatic stress disorder and one in which the same complex questions of mind and body, event and experience, past and present are raised. It is also critical for its role in the development of Freud's thought and hence for thinking through the relationship between psychoanalysis and trauma theory. Although Bergo does not phrase things in quite this way, I think it is one of the chief merits of her chapter that she is able to show the gradual shifts in Freud's understanding rather than attributing to him an abrupt reversal. For Bergo, Freud's own, possibly unconscious, motivations for shifting from a trauma etiology of hysteria to an oedipal model of psychic ontogeny have little to do with concealing or denying the "facts" of childhood sexual abuse. Rather, having received shocked disbelief for his early presentation of a case of male hysteria, and struggling in a cultural milieu that pathologized both women and Jews and rendered them ostensibly prone to hysteria, Freud shifted away from hysteria to repression (and hence, in Horowitz's terms, from the abnormal to his remarkable picture of the "normal" human condition).

I am not in a position to evaluate the historical details of the argument that, at least, would need to be supplemented by further attention to Freud's positive excitement about his new line of inquiry (which also faced hostility and skepticism from certain sectors of the public), but I do see a number of merits. For one thing, Bergo illustrates how Charcot and Freud point to the periodicity or discontinuity of symptoms, hence to body and mind as "not simply parallel entities . . . [but] interchangeable states" and thence to "Freud's subsequent disavowal of rigid distinctions between 'real' and 'imaginary' anxiety." Most significantly, events become traumas not immediately, but when recollections are triggered by subsequent events (*Nachträglichkeit*), a position that is quite different from much contemporary trauma theory. Second, Bergo shows how Freud continuously struggles intellectually with these issues, for instance with respect to comparing war-induced with sexual traumatism. However, Bergo also thinks that despite an original challenge to sexual and racial essentialism, an effect of Freud's later thinking was "a certain refeminization of hysteria and the erasure, or discrediting, of trauma-based hysteria in men."

If hysteria and even conversion hysteria (i.e., as manifest in bodily symptoms such as limb paralysis) remain present today, the question is how they are distinguished, in theory or in therapeutic practice, from trauma and traumatism. One difference in the way we conceive of the respective conditions concerns accountability. If hysteria is no longer determined by a person's gender or genetic inheritance, it still carries the suggestion of origins in the sufferer's psychological history, hence of the sort of weakness from which Freud, in Bergo's portrait, was trying to distinguish himself. Not every adult will show signs of hysteria. Trauma, by contrast, could happen to anyone; it is a consequence of accident. Hence its diagnosis carries no ostensible moral weight or negative judgment of the sufferer.[30] A second question that Bergo's chapter implicitly raises, then, is how contemporary cultural imaginaries of gender (or other qualities of personhood) play into both how we conceptualize these conditions, and how people fall into and repeat them.[31]

How literally we understand trauma may be related to how closely we encounter it. Obviously, critical theory and therapeutic action are different projects with different priorities and concerns. Horowitz notes the dissonance between the metapsychological and practical demands of psychoanalysis. How much greater then the dissonance between those who wish to theorize from trauma, those who wish to theorize about it, and those whose main endeavor is to offer therapy or advocacy to sufferers and hence whose first loyalty must be to them. Sandra Bloom and Judith Herman are well-known therapists and advocates of trauma therapy. Their relationship to the subject matter is necessarily different from that of philosophers or anthropologists and so I hope it will not be unfair for a disciplinary outsider to turn some of their arguments to close inspection.

Sandra Bloom's ambitious chapter includes some questionable anthropology and evolutionary assumptions. Moving beyond these—as well as the disturbing reification of trauma evident in her opening statistics—Bloom's chapter offers a masterly synthesis of much work on human distress, rewriting it in the key of "trauma," or even that of "stress" (a pair of concepts that all the contributors might take better care to distinguish). The analysis is particularly enhanced by the importation of cybernetic (systems) theory that enables Bloom to compare and connect individual and collective processes. Thus she argues that "groups designed to help people survive more successfully may end up becoming 'trauma-organized systems,' inadvertently organized around interactively repeating the patterns of repetition that are keeping the individuals they are serving from learning, growing, and changing." The idea that a constricted view of the world can come to be seen as normal is a powerful tool of critique.

Bloom is the only author here to write specifically of the prevailing social climate. For her, "over half of the population" of the United States are at risk of trauma and PTSD; one wonders what figures she would apply to the rest of the world—Iraqis, for example. I make this point not to criticize Bloom's ethical or political convictions but rather to underscore—with Bloom—Americans' sense of vulnerability despite the dissonance between the comforts of middle-class existence and the daily insecurity that appears to characterize the lives of the exceedingly large numbers of people elsewhere living in situations of low-intensity warfare, refugee camps, or ongoing nutritional deprivation. Bloom paints a picture of Americans as "survivors," caught in "traumatic reenactment." This is pathology on a mass scale—and, I would add—it is equally mass pathologization. Indeed, if PTSD is produced by mediatized "terror," one could say that reading Bloom's chapter could be dangerous to your health.

It becomes gradually apparent that Bloom is offering a political diagnosis of post-9/11 America and the pathologies of the Bush regime. She notes that "national mourning gave way to the drums of war and the grieving process was prematurely arrested and redirected in service of aggression." As Bloom says, "Paying attention only to physical safety does not make a group safe. . . . Real safety depends on at least three other domains: respect for individual rights, a shared sense of social responsibility, and a system of ethical conduct that is expressed as strongly in deeds as in words." This is a welcome intervention and well put. But the argument is somewhat reductive insofar as it relies exclusively on response to a single traumatic event (or its reenactment). After all, Bush's policies did not change overnight; many were in place long before 9/11 and economic interest was and remains significant. The elephant in the room belongs to the Republican Party.

It is worth remarking that Bloom is also the only contributor to recognize explicitly the challenge posed to all forms of psychotherapy by the shift to pharmacological intervention. To my knowledge, the philosophical challenges offered by the rise of psychopharmacology have barely begun to be addressed.[32]

Judith Herman begins with three assertions: much effort is given to silencing and denying atrocity; such effort is doomed to failure; "remembering and telling the truth about terrible events are essential tasks for both the healing of victims, perpetrators, and families and the restoration of social order." She cites psychological research on these topics but notes that we know little about the effects of their acts on perpetrators. For Herman, "disturbances of memory are a cardinal symptom of posttraumatic disorders" and she says strikingly about "traumatized people" that they "seem to have lost 'authority over their memories.'"

The latter remark is in line with many of the contributors, but Herman is perhaps the most succinct when it comes to outlining the condition: "The common denominator—the A criterion—of psychological trauma is the experience of terror." Citing the authority of a psychiatric textbook, she writes, "Traumatic events are those that produce 'intense fear, helplessness, loss of control, and threat of annihilation.' This is the definition in the fourth edition of the *Comprehensive Textbook of Psychiatry*, and extensive studies in the DSM-IV field trials have essentially confirmed this observation." This has the merits of specificity, but one cannot help asking how "field trials" can "confirm" what is actually a definition not an observation. Herman invokes and writes with the authority of science, yet it is by no means clear that science has privileged access to the domain of human suffering or that it is truly able to reduce it to a particular model of dysfunction.

I do not wish to deny that terrible experiences, especially those that occur early in life, cannot or do not produce miserable consequences, nor that "abnormalities of memory" may not be among these. I merely worry about the effects of such a scientifically authoritative and positive language that "produces the objects of which it speaks" and neglects to look at the way these objects become referentially self-confirming. To be sure, this worry is the role of the philosopher or anthropologist, not the clinician.

Herman is more interesting when she describes variation in forms of autobiographical memory and in noting that memories—she refers specifically to "traumatic" ones—"could manifest in disguised form as somatic and behavioral symptoms."[33] Herman argues that the goal of therapy is to "reintegrate memory," that is, to take experience that has been dissociated and shunted off into bodily or behavioral symptoms and transform it into words. This raises two main issues.

(1) In what sense do words "recover" and record "undisguised" a missing experience? Might they rather form a therapeutic interpretation of something in essence unknowable? If we are to speak of truth, is it a matter of certum—certainty, correspondence truth—or verum—meaningfulness, poetic truth? Remember, for Dobbs-Weinstein and perhaps for Beardsworth, the traumatic event is not consciously perceived. For Golden, "the traumatic wound . . . excludes linguistic representation," and "the retrospective trauma narrative . . . is incidental to the traumatic experience itself" (citing Gregg Horowitz). Brown critiques Herman to say, with respect to trauma, that speech is expressive of selfhood without necessarily being a form of representation. It is as though the nature of traumatic experience is such that it reduces the sufferer to the state of prereflective, presymbolic consciousness.[34] Any attempt to represent the trauma post facto is thus certain to be precisely not the thing that Herman wants it to be, that is, not an accurate representation (or reproduction) but a kind of performative speech act to which the criteria of truth and falsity in the correspondence sense do not apply. Acknowledging the illocutionary function of speech would place discussions of "false" memory syndrome in a new light and would change both the criteria by which we evaluate such narratives and the means by which we attempt to resolve both the conflicts from which they stem and the conflicts to which they give rise.

The inability to represent or accurately reproduce an event or experience means also that the ostensible silence of sufferers is not only a product of collusion or fear of being actively silenced by the perpetrators, but part of the very condition itself. Moreover, therapy in such a case must have as means or end something other than simply transforming "symptoms" into "words." This introduces the second main point.

(2) Is such a model of therapeutic healing exclusive? Is it universally appropriate? Horowitz suggests a very different answer when he says, "The central discovery in the psychoanalytic consulting room is that we do not cease suffering when we know the source of our suffering." Even without recourse to psychoanalytic irony, one can ask whether Herman is bound by a particular version of the mind-body problem.[35] Does healing always occur through discursive reason? Must the body be relegated to pathology ("disguised symptoms") and the mind to cure ("reintegration")? If transforming body into word may be therapeutic, could not also be the transformation of word or symbol into body? Or is this a false dichotomy? Studies of non-Western healers repeatedly demonstrate the positive effects of performative therapies.[36]

I could not agree more with Herman when she concludes, "The pursuit of truth in memory takes different forms in psychotherapy, where the purpose is to foster individual healing; in scientific research, where the purpose is to subject hypotheses to empirical test; and in court, where the purpose is to mete out justice. Each setting has a different set of rules and standards of evidence, and it is important not to confuse them." But Herman does, and we do.

Some people speak of trauma as though it were a single, known thing. I am not convinced. I share Dobbs-Weinstein's aim to "highlight the ambiguity of the term 'trauma.'" I note a lack of consensus on distinguishing the condition from its symptoms. There are also considerable differences in descriptions of the etiology. Is the trigger shock, stress, terror, violation of bodily boundaries, violation of trust? Do ostensibly similar events produce the same consequences? In fact, people do not all respond to the concentration camp, rape, or auto accidents in the same manner, suggesting that how an event is construed, both as and after it happens, does often retain—or regain—salience.[37] The indifference of trauma, suggests Scott, can also manifest as a kind of resilience.

Why have attributions of "trauma" (or PTSD) recently become so widespread in North America? Do they (or ought they to) refer to suffering in general, to a particular conception of suffering, to a particular kind of suffering, to a mode of expression? Does "trauma" refer to a source of suffering, to a symptom (or set of symptoms), to a disease or an illness?[38] Is it conceptualized as an original kind of event or act upon the person, a kind of experience of events or acts, a kind of personal (depersonalized) response to experience, or a kind of long-term condition in consequence of certain acts, experiences, or responses? Do we distinguish "actively suffering a traumatic experience" from "being traumatized," or from "having PTSD"? Is trauma a "thing" at all? Are we mistaking the lens or frame though which we see things for the thing itself? Conversely, can we talk about suffering in ways that avoid objectifying experience or labeling the sufferer? Is suffering fundamentally a biological condition—or a moral one?

What I would request is further consideration of trauma as a historically located phenomenon, that is, as a particular way of interpreting suffering rather than as the literal form or substance of suffering itself. Think for a moment about the very idea that suffering is something to be diagnosed rather than acknowledged, witnessed, or simply assuaged. What kind of society is it that cannot accept suffering without having to diagnose it? What does it mean to medicalize pain?[39]

The question then becomes: how are experiences of suffering shaped and articulated by being defined and described—even diagnosed and treated—as trauma? What difference does it make to suffering and for sufferers to describe their condition as trauma?[40] What difference does it make to society and for politics? To so describe suffering is to reify and medicalize it and to participate in those larger processes characteristic of the present moment of history in which the political is reduced to the individual and in which the mind-body distinction has been overtaken by a distinction between my self and my mind or body, in which even our minds become objects to ourselves, and to our physicians.

The risks of not taking up a trauma model are well described and argued by Sandra Bloom and Judith Herman. The risks of taking up the trauma model, and especially of accepting it to the exclusion of other models, are less explicitly described here.[41] While the concept of trauma raises lively thought among philosophers concerning issues of mind and body, subjectivity and experience, the cumulative import of these chapters is that one risk of holding too close to the concept of trauma is an oversimplification of human suffering and action. This has several sources.

A first source lies in ignoring the significance of culture, both in mediating the experience of events—indeed, in construing what an "event" is—and in addressing the effects of violation and providing the means for regeneration. If the cultural construal of meaning is relevant, to what degree can we generalize about specific kinds of events as "traumatic" or about specific modes of accountability, and to what degree ought we to refrain from assuming, propagating, or exporting an ostensibly universal model? For example, preliminary research carried out in the aftermath of the brutal war in Sierra Leone suggests that (embodied) ritual purification may be more salient and effective for personal healing and social reconciliation than (discursive) disclosure and confession.[42]

A second oversimplification, one that these chapters go a long way to address, lies in philosophical naiveté and specifically in tacit attachment to a specific version of the mind-body problem without adequate consideration of the alternatives, such as the nondualist position put forward by Kristen Golden.[43] An additional philosophical question raised in many of the chapters concerns the nature of representation. Not only does trauma raise questions about how experience is apprehended by those who suffer it, but it provokes ethical quandaries about how such suffering is in turn to be represented to others in art, the public media, or theory. The biology of terror cannot be disarticulated from the moral investments of sufferers, perpetrators, and witnesses, including, at times, the moral investment in silence.

A third source of oversimplification lies in ignoring Freudian insights about attachment, conflict, fantasy, identification, suggestion, and ambivalence—and more recent psychoanalytic accounts of human relationality (object relations theory no less than Kristeva). The arguments presented in this volume have not reached (nor tried to reach) a consensus that trauma is a single, discrete, unitary phenomenon. They challenge us to be skeptical of an approach that ignores loss or that takes trauma too literally, without some measure of irony about what we can know or how we relate to the world.

A fourth simplification, appropriately flagged by Bloom, reduces collective suffering and accountability to the individual, neglecting the social milieu and the forms of structural violence that enable trauma. Depoliticization is a problematic "side effect" of all forms of medicalization. Conversely, promoting a trauma model as an alternative to orthodox Freudianism has been a specifically political intervention with respect to sexual and domestic abuse. The situation now is that any explicit position taken with respect to the prevalence or significance of trauma is going to be understood as politically motivated and bearing consequences.

This leads to my final form of simplification, namely, ignoring Foucauldian insights concerning discursive regimes of power and especially the very power to classify and categorize experience. How do forms of psychotherapy reproduce and produce particular forms of selfhood and implicit cultural assumptions about persons, relationships, and experience? How do they work to open certain fields of cognitive discrimination—and thereby shut off others? What can we say about the genealogy of trauma, about the historical ontology of ourselves? To what degree or in what ways might the discourse of trauma produce the very kinds of human subjects it presupposes? It would thereby approximate a closed, self-fulfilling system of thought, much as Evans-Pritchard described witchcraft and its diagnosis among the Azande.[44]

Philosophically, what is needed is the sort of principled skepticism displayed by Ian Hacking. Hacking is not a full social constructionist. Child abuse, as Hacking affirms, is real. So is terror. Hacking makes a useful distinction between what he calls indifferent and interactive kinds. Of interactive kinds Hacking says, "What was known about people of a kind may become false because people of that kind have changed in virtue of how they have been classified, what they believe about themselves, or because of how they have been treated as so classified. There is a looping effect." Conversely, indifferent kinds do not change "because they become aware of what we know."[45] I question the implicit assumption that trauma (qua condition) is an indifferent kind. Not to say that this is necessarily wrong, only that the alternative remains underexamined.

Foucauldians show how within certain historically located and continuously changing discursive structures and disciplinary practices people in all their complexity get reduced to kinds. These come to look exclusive and definitive, based on a binary bureaucratic logic—either you are one of those kinds of people—a hysteric, a criminal, a child abuser, a trauma victim—

or you are not. They also get modified and more powerfully legitimated by statistical understanding with the production of norms and deviance. The looping effect—the interaction between the category and what it denotes—is inevitably overlooked by practitioners. That is how a certain kind of discursive regime works—without qualification, without contextualization, without uncertainty, without irony. But that is not really how people are, is it?

Hence we need theories that can address both the production of kinds and their impact on subjectivity and the nature of human subjectivity itself and the way it responds not only to particular kinds of violent acts or disturbing experiences but also to attributions of kindedness or particular classificatory schemes and to investments in or habituations to particular logics of practice. How do people integrate or resist being kinded?[46]

Suffering is everywhere; so is silence; so too is the proliferation of discourse. Discourses of suffering are not in and of themselves good or bad things. As in Foucault's formulation of sexuality, they produce extensive knowledge and this knowledge in turn affects (and sometimes effects) its objects as well as its subjects. The things to ask about discourses are how they proliferate and what their effects are. They need to be located within wider nexuses of tradition, institutions, history, and politics. A discourse of "trauma" makes little sense without locating it with respect to a specific time and place; millennial America, for instance, with its particular configuration of gender, race, and class, its postpuritan religious ethos, readiness for litigation, military adventurism, possessive individualism, scientism, fear-mongering media, anxieties over consumption and downward mobility, and so forth. The point is not that either trauma (the concept) or trauma (the experience) is directly produced by any of these forces or conditions specifically, but rather that the elaboration of particular discursive practices occurs in relation to them.

To recognize these forms of simplification is also to make a call for more ethnography. Ethnography can demonstrate the way that diagnoses of trauma fit particular narratives and acts of accountability and blame and the needs and interests of particular kinds of institutions or discourses. Trauma lays out unambiguous victims and villains; it thus serves as a morality play, to revisit Evans-Pritchard, a kind of local and contemporary witchcraft scenario.[47] This is one of trauma's attractions and one explanation for its contagiousness.

Ethnography would demonstrate additional paths of discursive mobility: how trauma is deployed in psychiatric diagnosis; how it moves between therapy, law, science, social work, intellectual movements, and the news media; how it is taken up in the lives of sufferers; how it articulates with alternative diagnoses such as depression and with the pharmaceutical industry and insurance regimes; how it draws on metaphoric connections between the nation and the individual (connections variously noted by Beardsworth, Bergo, and Bloom). We need exploration of how the state sets the terms of suffering. And finally, we need more ethnography on the suffering of ordinary people. What is the aftermath of a war or natural disaster like for survivors?[48] How do people and communities actually continue to live in the wake of terror?

Acknowledgments

Sincere thanks to the editors for the opportunity to engage with such stimulating chapters and for their close responses to a first draft of this chapter; they have improved it substantially but of course are not responsible for its weaknesses. Thanks also to Nadia Lambek for sharing her undergraduate paper on the aftermath of the Cambodian genocide, to Simon Lambek for lending me his copy of *King Oedipus*, and to Paul Antze for continuing to instruct me in the fields of psychoanalysis and trauma studies.

Notes

The (nonconsecutive) passages in the chapter's opening epigraph are from *The Theban Plays*, trans. E. F. Watling (Harmondsworth, UK: Penguin, 1974), 62, 63, and 67, respectively.

1. The phrasing and conceptualization is that of Marshall Sahlins in "The Original Affluent Society," *Stone Age Economics* (Chicago: Aldine, 1972). For a cogent overview of the debates to which his original formulation has given rise, see Jacqueline Solway, "The Original Affluent Society: Four Decades On," in *Egalitarian Politics: Anthropological Essays in Honour of Richard B. Lee*, ed. J. Solway (New York: Berghahn, 2006). She reprints an abridged version of Sahlins's essay in the same volume.

2. This term has been developed by Veena Das, Arthur Kleinman and others. See ed. A. Kleinman et al., *Social Suffering* (Berkeley and Los Angeles: University of California Press, 1997); and ed. Veena Das et al., *Violence, Social Suffering, and Recovery* (Berkeley and Los Angeles: University of California Press, 2001).

3. On this point, see David Healy, *Let Them Eat Prozac: The Unhealthy Relationship between the Pharmaceutical Industry and Depression* (New York: New York University Press, 2004).

4. Davos! Mann could not have imagined the subsequent irony entailed by his locale. Thomas Mann, *The Magic Mountain* (1924), trans. John E. Woods (New York: Knopf, 1995).

5. See, for example, Michel Foucault, *The History of Sexuality, Volume 1* (New York: Vintage, 1980); Ian Hacking, *The Social Construction of What?* (Cambridge, MA: Harvard University Press, 1999), especially the chapter "Kind-Making: The Case of Child Abuse," 125–62. See also Ian Hacking, "Memory Sciences, Memory Politics," in *Tense Past: Cultural Essays in Trauma and Memory*, ed. Paul Antze and Michael Lambek (New York: Routledge, 1996), 67–87; and Ian Hacking, *Rewriting the Soul: Multiple Personality and the Sciences of Memory* (Princeton, NJ: Princeton University Press, 1995). On trauma and PTSD, see respectively, Ruth Leys, *Trauma: A Genealogy* (Chicago: University of Chicago Press, 2000), hereafter cited in the text as *TG*, and Allan Young, *The Harmony of Illusions: Inventing Post-Traumatic Stress Disorder* (Princeton, NJ: Princeton University Press, 1995). These last two books are major works on the subject.

6. I do not wish to suggest that understanding spirit possession has been straightforward, nor is it an easy phenomenon to describe satisfactorily in a few words here. See Michael Lambek's *Human Spirits: A Cultural Account of Trance in Mayotte* (Cambridge: Cambridge University Press, 1981), *Knowledge and Practice in Mayotte: Local Discourses of Islam, Sorcery, and Spirit Possession* (Toronto: University of Toronto Press, 1993), and *The Weight of the Past: Living with History in Mahajanga, Madagascar* (New York: Palgrave-Macmillan, 2002). Spirit possession in a variety of forms is actually widespread in human societies. For an exemplary analysis see Janice Boddy, *Wombs and Alien Spirits: Women, Men, and the Zar Cult in Northern Sudan* (Madison: University of Wisconsin Press, 1989); and for a bibliographic and theoretical guide, see her "Spirit Possession Revisited: Beyond Instrumentality," *Annual Review of Anthropology* 23 (1994): 407–34.

7. Leys, *TG*.

8. Conversely, sufferers may not so easily move from "presentation" to "representation" that has been, since Freud, a goal of therapy.

9. For a compelling account of life by an anthropologist with a tumor growing up his spine, see Robert Murphy, *The Body Silent* (New York: Henry Holt, 1987).

10. This point is brought out in the various nonfunctionalist and generally European redactions of Freudian theory. For a recent American example, see Jonathan Lear, *Happiness, Death, and the Remainder of Life* (Cambridge, MA: Harvard University Press, 2000), hereafter cited in the text as *HDL*. For a concise and elegant essay on Freud and memory that speaks to many of the issues discussed here, see Paul Antze, "The Other Inside: Memory as Metaphor in Psychoanalysis," in *Regimes of Memory*, ed. Susannah Radstone and Katharine Hodgkin (London: Routledge, 2003), 96–113.

11. See, for example, Michael Silverstein. "Shifters, Linguistic Categories, and Cultural Description," in *Meaning in Anthropology*, ed. Keith Basso and Henry Selby (Albuquerque: University of New Mexico Press,

1976). A very interesting account of ritual from the perspective both of communication and human adaptation is Roy A. Rappaport, *Ritual and Religion in the Making of Humanity* (Cambridge: Cambridge University Press, 1999), hereafter cited in the text as *RR*. An earlier attempt to address some of the mind-body issues posed by communications theory is Gregory Bateson, *Steps to an Ecology of Mind* (New York: Ballantine, 1972). There is also, of course, the whole field of semiotics dedicated to addressing such questions. For an entry to the field, see the journal *Semiotica*.

12. Consider on indispensability, George Lakoff and Mark Johnson, *Metaphors We Live By* (Chicago: Chicago University Press, 1980); on harm, both Susan Sontag, *Illness as Metaphor* (New York: Vintage, 1978) and Emily Martin, *The Woman in the Body* (Boston, MA: Beacon, 2001). Allan Young argues specifically that the migration of "trauma" from surgical to psychological shock was not one of analogy. See his "Bodily Memory and Traumatic Memory," in *Tense Past: Cultural Essays in Trauma and Memory*, ed. Paul Antze and Michael Lambek (New York: Routledge, 1996), 89–102.

13. On habitus, see Pierre Bourdieu, *Outline of a Theory of Practice* (Cambridge: Cambridge University Press, 1977) and, before him, Marcel Mauss, "Techniques of the Body" (1935), *Economy and Society* 2 (1973): 70–88. On memory from this perspective, see Paul Connerton, *How Societies Remember* (Cambridge: Cambridge University Press, 1989). Substantive philosophical and ethnographic discussions of place can be found, respectively, in Edward Casey, *Getting Back into Place: Toward a Renewed Understanding of the Place-World* (Bloomington: Indiana University Press, 1993) and Keith Basso, *Wisdom Sits in Places* (Albuquerque: University of New Mexico Press, 1996). An emerging "site" where habitus and mind-body communication issues of the kind raised by Golden with respect to trauma are to be found is autism.

14. My understanding of irony—"Irony and Illness: Recognition and Refusal" and "Rheumatic Irony: Questions of Agency and Self-Deception as Refracted through the Art of Living with Spirits," both in *Illness and Irony: On the Ambiguity of Suffering in Culture*, ed. Michael Lambek and Paul Antze (New York: Berghahn, 2004), 1–19 and 40–59—is highly indebted to Alexander Nehamas, *The Art of Living: Socratic Reflections from Plato to Foucault* (Berkeley and Los Angeles: University of California Press, 1998) as well as to Kenneth Burke, "Four Master Tropes," *A Grammar of Motives* (New York: Prentice Hall, 1945) 503–17. Mind-body understood in this sense is prefigured in the medical anthropological distinction between "illness" and "disease." See Leon Eisenberg, "Disease and Illness," *Culture, Medicine and Psychiatry* 1 (1977): 9–23; Arthur Kleinman, *Patients and Healers in the Context of Culture* (Berkeley and Los Angeles: University of California Press, 1980); and Arthur Kleinman, "Editor's Note,"

Culture, Medicine and Psychiatry 7 (1983): 97–99. A similarly stubborn opposition is that between "sex" and "gender." See Judith Butler, *Gender Trouble* (New York: Routledge, 1990); and Rita Astuti, "'It's a Boy,' 'It's a Girl!': Reflections on Sex and Gender in Madagascar and Beyond," in *Bodies and Persons: Comparative Perspectives from Africa and Melanesia*, ed. M. Lambek and Andrew Strathern (Cambridge: Cambridge University Press, 1998), 29–52. On irony in Freud, see Paul Antze, "Illness as Irony in Psychoanalysis," in *Illness and Irony: On the Ambiguity of Suffering in Culture*, ed. Michael Lambek and Paul Antze (New York: Berghahn, 2004), 102–22. The position I describe is also commensurate with Ruth Leys's (*TG*) conclusions concerning the "unresolved" (305) and "unresolvable" (307) tension between mimesis and antimimesis.

15. Michael Lambek, "Body and Mind in Mind, Body and Mind in Body: Some Anthropological Interventions in a Long Conversation," in *Bodies and Persons: Comparative Perspectives from Africa and Melanesia*, ed. M. Lambek and Andrew Strathern (Cambridge: Cambridge University Press, 1998), 103–23.

16. The distinction between *certum* and *verum* is from Giambattista Vico, *The New Science*, trans. Bergin and Fisch (Ithaca, NY: Cornell University Press, 1948). I have drawn from Rappaport, *RR*.

17. For a very influential dialectical account, see Peter Berger and Thomas Luckmann, *The Social Construction of Reality* (Garden City, NY: Anchor Books, 1966). See also Michael Lambek, "The Past Imperfect: Remembering as Moral Practice," in *Tense Past: Cultural Essays in Trauma and Memory*, ed. Paul Antze and Michael Lambek (New York: Routledge, 1996), 235–54.

18. See Lambek as cited in note 15. I need to be very clear that in describing trauma patients as suffering from literalism I am taking a contrary position to the literalism ascribed by Leys (*TG*) to trauma theorists such as Bessel Van der Kolk and Cathy Caruth who see symptoms as literal repetitions of the trauma. My point is not that the symptoms are literal but that sufferers may be able to understand them in only that fashion. This is not to take a position either way in regard to the facticity of specific events or the accuracy of specific memories. On the distinction between the veridical and the literal and their elision by Van der Kolk and Caruth, see Leys, *TG* 229.

19. Clifford Geertz, "The Growth of Culture and the Evolution of Mind" and "The Impact of the Concept of Culture on the Concept of Man," *The Interpretation of Cultures* (New York: Basic Books, 1973), 33–83.

20. Marshall Sahlins, *Culture and Practical Reason* (Chicago: University of Chicago Press, 1976).

21. For a Merleau-Pontian account of the limits of culture from an anthropological perspective, see Thomas Csordas, ed., *Embodiment and*

Experience: The Existential Ground of Culture and Self (Cambridge: Cambridge University Press, 1994).

22. To mention only two particularly rich examples, see James Fernandez, *Bwiti: An Ethnography of the Religious Imagination in Africa* (Princeton, NJ: Princeton University Press, 1983), and Gananath Obeyesekere, *Medusa's Hair: An Essay on Personal Symbols and Religious Experience* (Chicago: University of Chicago Press, 1981).

23. See Lear, *HDL*.

24. Frederick Crews, in particular, used the weakness of the position of the advocates of "recovered memory" as a weapon with which to attack psychoanalysis, despite the fact that these advocates were, for the most part, not psychoanalysts. See his *The Memory Wars: Freud's Legacy in Dispute* (New York: New York Review of Books, 1995). In defense of Freud, see Jonathan Lear, "The Shrink Is In," *New Republic* (December 25, 1995), 18–25.

25. For a recent review of some of the debate as well as an excellent synthesis of the latest research on trauma and memory from diverse branches of academic psychology, see Richard J. McNally, *Remembering Trauma* (Cambridge, MA: Belknap/Harvard University Press, 2003), hereafter cited in the text as *RT*.

26. For an extensive and sophisticated discussion of the complex relationships between psychoanalysis and trauma theory see Leys, *TG*. Leys's central theme concerns the tension between mimetic and antimimetic tendencies in theorizing trauma, between emphasizing the significance of imitation, suggestion, and repetition, on the one hand, and the impingement of the external event on the autonomous subject, on the other hand. Curiously, questions of mimesis and suggestion are not explicitly taken up by any of the contributors to the present volume.

27. This memorable phrase is from Leonard Shengold, *Soul Murder Revisited: Thoughts about Therapy, Hate, Love, and Memory* (New Haven, CT: Yale University Press, 1999).

28. While revising this chapter five months after the tsunami, I did not encounter reports of any follow-up studies in the popular media.

29. For an account of why the suffering of Cambodian refugees is better understood as cultural bereavement than as PTSD, see Maurice Eisenbruch, "From Post-Traumatic Stress Disorder to Cultural Bereavement: Diagnosis of Southeast Asian Refugees," *Social Science and Medicine* 34, no. 6 (1991): 673–80,

30. Indeed, in the case of American soldiers, Allan Young ("Bodily Memory and Traumatic Memory") argues that its function has been to absolve perpetrators of accountability.

31. I cannot address the historicization of gendered illness, moral personhood, and cultural imaginaries here but suggest that in addition to Sander Gilman (as cited by Bergo), the work of Susan Bordo—*Unbearable Weight:*

Feminism, Western Culture, and the Body (Berkeley and Los Angeles: University of California Press, 1995)—is a good place to start.

32. For an illustration, see Andrew Lakoff, "The Lacan Ward: Pharmacology and Subjectivity in Buenos Aires," in *Bodies and Persons: Comparative Perspectives from Africa and Melanesia*, ed. M. Lambek and Andrew Strathern (Cambridge: Cambridge University Press, 1998), 82–101. See also Tanya Luhrmann, *Of Two Minds: The Growing Disorder in American Psychiatry* (New York: Knopf, 2000).

33. A question to ask, not of Herman, but of the philosophers who reject dualism, is whether they would go along with considering the somatic and behavioral manifestations as "disguised" or, for that matter, as "symptoms."

34. All of this may be moot if we accept the result of clinical research that, "[e]vents that trigger overwhelming terror are memorable, unless they occur in the first year or two of life or the victim suffers brain damage. The notion that the mind protects itself by repressing or dissociating memories of trauma, rendering them inaccessible to awareness, is a piece of psychiatric folklore devoid of convincing empirical support" (McNally, *RT* 275).

35. A lot here rests on what psychoanalysts mean by "insight."

36. A famous early example is Claude Lévi-Strauss, "The Effectiveness of Symbols" (1949), in *Structural Anthropology*, vol. 1 (New York: Basic Books, 1963). See also Victor Turner, *The Forest of Symbols* (Ithaca, NY: Cornell University Press, 1967); René Devisch, *Weaving the Threads of Life* (Chicago: University of Chicago Press, 1993); and Lambek, *Knowledge and Practice in Mayotte.*

37. For example, McNally emphasizes the effects of guilt or shame in the initial experience and notes, "That guilt about having harmed others can produce PTSD underscores the moral complexity of trauma and the limitations of animal models of PTSD" (*RT* 85).

38. Beardsworth helpfully notes that for Freud "the event is not traumatic but traumatogenic." Similarly, Bergo cites Didier Anzieu's distinction between "trauma as a trigger and traumatism as an effect, physical or psychological, on the victim." On the disease/illness distinction, see note 15.

39. Compare, for example, Paul Ricoeur, *The Symbolism of Evil* (Boston, MA: Beacon, 1967).

40. For example, whereas PTSD focuses on symptoms, cultural bereavement attends to meanings (Eisenbruch, "From Post-Traumatic Stress Disorder to Cultural Bereavement" 676).

41. See especially the work of Ruth Leys and Allan Young referred to in note 5.

42. I refer to the very interesting work of Rosalind Shaw, much of which is still forthcoming, but see "Rethinking Truth and Reconciliation Commissions: Lessons from Sierra Leone," *Special Report for the United States*

Institute of Peace, no. 130 (February 2005) See also Michael Jackson's deliberate evocation of W. G. Sebald's elegiac mix of prose and photography, *In Sierra Leone* (Durham, NC: Duke University Press, 2004).

43. See her *Nietzsche and Embodiment: Discerning Bodies and Non-dualism* (Albany: State University of New York Press, 2006).

44. E. E. Evans-Pritchard, *Witchcraft, Oracles, and Magic among the Azande* (Oxford: Clarendon Press, 1937).

45. Ian Hacking, "Madness: Biological or Constructed?" in *The Social Construction of What?* (Cambridge, MA: Harvard University Press, 1999), 104, 105.

46. Phrased in this manner the questions are not dissimilar to those concerning gender posed in the feminist literature. For a major intervention that moves the argument definitively beyond questions of resistance, see Saba Mahmood, *Politics of Piety: The Islamic Revival and the Feminist Subject* (Princeton, NJ: Princeton University Press, 2005).

47. See Jean LaFontaine, *Speak of the Devil: Allegations of Satanic Child Abuse in Contemporary England* (Cambridge: Cambridge University Press, 1998). Also Jean Comaroff, "Consuming Passions: Child Abuse, Fetishism, and 'The New World Order,'" followed by my response, Michael Lambek, "Monstrous Desires and Moral Disquiet," *Culture* 17 (1997): 19–25.

48. On the making of PTSD, see Young, *The Harmony of Illusions*; on the state and torture, see Talal Asad, *Formations of the Secular: Christianity, Islam, Modernity* (Stanford, CA: Stanford University Press, 2003), and Allan Feldman, "Strange Fruit: The South African Truth Commission and the Demonic Economies of Violence," in *Beyond Rationalism: Rethinking Magic, Witchcraft, and Sorcery*, ed. Bruce Kapferer (New York: Berghahn, 2002), 234–65. Judith Zur—*Violent Memories: Mayan War Widows in Guatemala* (Boulder, CO: Westview, 1998)—and Linda Green—*Fear as a Way of Life: Mayan Widows in Rural Guatemala* (New York: Columbia Univ. Press, 1999)—provide accounts of life after the genocidal action in that country, while Veena Das—*Life and Words: Violence and the Descent into the Ordinary* (Berkeley and Los Angeles: University of California Press, 2006)—is a particularly subtle analysis of the aftermath of violence in India.

Contributors

SARA BEARDSWORTH is associate professor in the Philosophy Department at Southern Illinois University. She is author of *Julia Kristeva: Psychoanalysis and Modernity* (2004). Her research is in nineteenth- and twentieth-century European philosophy, and she has published articles on psychoanalysis, feminism, and Frankfurt School critical theory.

BETTINA BERGO is associate professor of philosophy at the Université de Montréal. Author of *Levinas between Ethics and Politics* (2002), she has translated three of Levinas's works; M. Zarader's *The Unthought Debt: Heidegger and the Hebraic Heritage* (2006), *Judeities: Questions to Jacques Derrida* (2007), and *Dis-Enclosure: Deconstruction of Christianity* (2008)—the last two with Michael B. Smith. She is currently working on the translation of Didier Franck's *Nietzsche and the Shadow of God*. The author has written articles on phenomenology, psychoanalysis, feminism, and critical race theory, and she is writing a brief history of anxiety in twentieth-century thought.

SANDRA L. BLOOM, M.D., is a board-certified psychiatrist, 2005 recipient of Temple University School of Medicine's Alumni Achievement Award, and executive director of Community Works Inc, an organizational consulting firm committed to the development of nonviolent environments. Dr. Bloom is the founder of the Sanctuary Model, a trauma-sensitive approach to residential treatment programs for adults and children and the Sanctuary Institute. She currently serves as associate professor of health management and policy at the School of Public Health at Drexel University in Philadelphia, Pennsylvania, and is Distinguished Fellow of the Andrus Children's Center in Yonkers, New York.

IDIT DOBBS-WEINSTEIN, associate professor of philosophy at Vanderbilt University, is the author of *Maimonides and St. Thomas on the Limits of Reason* (1995) and *Moses Maimonides and Medieval Jewish Philosophy* (1996). She is coeditor with Lenn E. Goodman and James A. Grady of the forthcoming *Maimonides and His Heritage* (2009). Professor Dobbs-Weinstein's ongoing research project seeks to retrieve a Materialist Aristotelian tradition occluded by theologico-political forces. Her articles address the influence of Medieval Jewish and Islamic philosophers upon subsequent materialists spanning from Spinoza, through Marx and Freud, to Benjamin and Adorno.

MICHAEL L. GALATY is associate professor of anthropology at Millsaps College. He is the editor (with Charles Watkinson) of *Archaeology under Dictatorship* (2004). Together with Albanian colleagues, he directs the Shala Valley Project, an archaeological and ethno-historic survey of a northern Albanian tribal region (www.millsaps.edu/svp).

KRISTEN BROWN GOLDEN is associate professor of philosophy at Millsaps College. Her publications include *Nietzsche and Embodiment: Discerning Bodies and Non-dualism* (2006) and journal articles on social and political topics in Aristotle, Nietzsche, and Merleau-Ponty. Brown orients her perspective through nineteenth- and twentieth-century European philosophy. She is currently researching themes in psychoanalytic, clinical, and scientific discouses of trauma and their nodes of intersection with race theory.

JUDITH LEWIS HERMAN, M.D., is clinical professor of psychiatry at Harvard Medical School and director of training at the Victims of Violence Program at the Cambridge Hospital. She is the author of two award-winning books: *Father-Daughter Incest* (1981) and *Trauma and Recovery* (1992). Herman is recipient of the 1996 Lifetime Achievement Award from the International Society for Traumatic Stress Studies and the 2000 Woman in Science Award from the American Medical Women's Association. In 2007 she was named a Distinguished Life Fellow of the American Psychiatric Association.

GREGG M. HOROWITZ is associate professor of philosophy at Vanderbilt University. He is the author of *Sustaining Loss: Art and Mournful Life* (2001). He is also coeditor, along with Tom Huhn, of *The Wake of Art: Criticism, Philosophy, and the Ends of Taste* (1998). In 2008 he was the Berthold Leibinger Fellow at the American Academy in Berlin, where he was working on a book on psychoanalysis and modern politics, and writing, piecemeal, on the fate of archaic media in contemporary art.

MICHAEL LAMBEK is professor of anthropology at the University of Toronto; from January 2006 through July 2008, he was also a professor in the Department of Anthropology at the London School of Economics. He is the author of three ethnographic works on Malagasy speakers of the western Indian Ocean, focusing in part on interpreting their systems of spirit possession, and is the editor of a number of other books, including *Tense Past: Cultural Essays in Trauma and Memory* (1996) and *Illness and Irony: On the Ambiguity of Suffering in Culture* (2004), both coedited with Paul Antze.

ERIC SEAN NELSON is assistant professor of philosophy at the University of Massachusetts, Lowell. He has published articles on issues in ethics, philosophy of religion, and social-political theory in Kant, Schleiermacher, Dilthey, Adorno, Heidegger, and Levinas. He is the coeditor of *Addressing Levinas* (2005) and *Rethinking Facticity* (2008).

CHARLES E. SCOTT holds the positions of Distinguished Professor of Philosophy and director of the Vanderbilt University Center for Ethics at Vanderbilt University. His most recent books include *The Time of Memory* (1999), *Lives of Things* (2002), and *Living with Indifference* (2007).

SHARON R. STOCKER directs a team that is currently studying unpublished artifacts from the excavations of the Palace of Nestor in Greece. In Albania she was codirector of regional surveys at Durrës and Apollonia, and now codirects excavations of a classical temple near Apollonia. Recent publications include "Archaeological Survey in the Territory of Epidamnus/Dyrrachium (Albania)"; "Deriziotis Aloni: A Small Prehistoric Site in Messenia"; and "Animal Sacrifice, Archives, and Feasting at the Palace of Nestor."

CHARLES WATKINSON is director of publications at the American School of Classical Studies at Athens. An archaeologist by training, he continues to be involved in fieldwork in Albania and Greece. He is the editor, with Michael L. Galaty, of *Archaeology under Dictatorship* (2004).

Index

Printed in Great Britain
by Amazon.co.uk, Ltd.,
Marston Gate.